This study was led by the directors of Energy Process Developments Ltd,
Dr. Trevor Griffiths, Jasper Tomlinson and Rory O'Sullivan (also Project Manager).

Along with guidance throughout the study, the report has been reviewed by the supervisory
panel, Professor Derek Fray FRS and Dr Geoff Parks of University of Cambridge and
Professor Paul Madden FRS of University of Oxford.

Major contributions have been provided by Frazer-Nash Consultancy, David Glazbrook and
Dr. Andrei Horvat.

The MSR community has been very cooperative and thanks are due to the teams at
Flibe Energy, ThorCon Power, Moltex Energy, Seaborg Technologies, Terrestrial Energy and
Transatomic Power along with:
the advanced reactor team at Oak Ridge National Laboratory, Atkins Ltd, Copenhagen
Atomics, International Thorium Molten-Salt Forum of Japan, Arthur Berenfeld, Clive
Elsworth, Charles Forsberg, Professor Bernard Gibbs, Robert Hargraves, Andy Kiang, Steve
Lyons, Ian Maciulis, Chad Manian and Barry Snelson MBE.

"When the facts change, I change my mind. What do you do, sir?"

\- John Maynard Keynes

MSR Design Proposers:

This report is available at www.EnergyProcessDevelopments.com

Executive Summary

It is widely accepted that the safe harnessing of energy from nuclear fission is a necessary component of a rational and sustainable energy policy. A central concern for the feasibility study reported here is the problem of finding the most suitable way of effectively and safely doing this. Liquid-fuelled molten salt reactors have been recognised as an excellent solution. China alone has initiated a major programme to pursue this opportunity. Past reviews have concluded that MSRs are many years away from implementation. The study undertaken for this report indicates that, following a decade of work, several small to medium developers - without need for more science - claim they are ready now with proposals for the next step to implementation, namely engineering design to prepare the safety case and to proceed to design and build. Six specific proposals have been reviewed for this study. These proposal assessments are the core substance of this study, with one proposal identified for development in the UK, the Stable Salt Reactor.

This study originated with a concern that current nuclear new build projects appear to be locked into the original solid-fuelled reactor technology. Since the 1970s the industry has lacked innovation. By increasing regulation and subsequent cost the result is an expensive energy source. The proposals considered for this study are for inherently safe efficient liquid-fuelled reactors which have the potential to be engineered to compete with fossil fuel prices. This solution needs to be conveyed with the help of this report to interested members of the public, institutions, the media, and to decision makers both in Government and in industry.

The opportunity to carry out this study owes a lot to Innovate UK funding and to voluntary contributions from individual engineers, consultancies and academics. An opinion poll carried out for this study helped identify public concerns and aspirations of those supporting more nuclear power. The media and institutions have been involved where good relations have developed. The team has been invited to present the progress of the study across the UK and internationally.

The team that has been engaged in this study has included, in addition to the three active directors of Energy Process Developments Ltd, several individual well-equipped engineers and support staff and expertise from engineering enterprises with leading positions in the nuclear industry, together with a supervisory panel of three distinguished academics.

The major obstacle to necessarily long-term plans for implementation of innovative nuclear reactor projects is funding. Large amounts of investment are needed, measured in hundreds of millions of pounds for first-of-a-kind start-ups of nuclear devices. In the initial stage of such a project, industry is not expected to take a lead, rather to follow the investment of public funds. After overcoming this first hurdle, hopefully in the lifetime of the present government, steps to industrial application will be undertaken. Academia can develop a collaborative programme to build a comprehensive basis of knowledge and expertise. This sector, already scarred from past events, cannot afford future failures. The investment, in the tens of billions of pounds – increasingly from industry – can establish a new face to nuclear with a world class industry-standard nuclear reactor system. The reward, apart from effectively addressing energy poverty both at home and abroad, is a stake in a nuclear power market estimated at a trillion pounds.

The authors of this report recommend to all who are interested that they should make the urgent necessary investment and commitment to an agenda to proceed with a molten salt reactor programme including a demonstration prototype as identified by this study.

Contents

1 Opportunities & Industry Overview

A historic decline of R&D and other investment in UK civil nuclear power is described. Current policy is directed to improvements in solid-fuelled reactors, including small to medium sized versions, together with emphasis on researching nuclear fusion possibilities. Liquid-fuelled reactor technology has been perceived as too far in the future to justify current attention.

Constraints in the UK nuclear industry sector include human resources availability and adequate investment. As an outcome what is proposed as policy may not be achievable. The suitability of more-of-the-same solid fuelled PWR technology to provide for planned increases of nuclear power in the energy sector is questioned.

Financial and economic returns from liquid-fuelled MSR technology are described; potential environmental benefits in terms of global warming are outlined.

2 MSR Designs Assessed in this Study

Criteria and methods for ascertaining availability of valid MSR proposals for a UK demonstration reactor are explained. Outlines of six alternative proposals are presented. Each individual outline indicates relevant features. Five out of the six designs are directly descended from the 1960s experimental reactor (MSRE). One proposal – the Moltex Stable Salt Reactor – is a development of an earlier concept from the same MSRE group.

3 Historical Background

Nuclear fission liquid-fuelled reactors originated as a conjecture at the time when atomic weapons were under development in the 1940s. Insights from that time onwards were brought to reality in the late 1960s by Alvin Weinberg who became director of the Oak Ridge National Laboratory. Civilian nuclear power was initiated in the USA with the Shippingport pressurised water reactor, a land-based adaptation of PWR technology chosen for the Nautilus, the first US Navy nuclear submarine. Further development of MSR technology in the US and elsewhere was effectively abandoned in 1976. The concepts were kept alive sufficiently to allow resurrection in the last two decades.

4 An Introduction to Liquid-fuelled MSR Technology

Neutrons, electrically neutral particles, together with electrons and protons which carry equal but opposite electric charges, are the principal components of atoms. Neutron activity induces nuclear fission which in a fuel salt provides the means for harnessing the energy locked into component atoms. Neutronic interactions with design materials and fuel salts are described.

5 MSR Benefits

Informed public opinion when polled showed concern about eight specific topics. The benefits derived from generic MSR technology when addressing these topics - and some others - are reviewed one by one. In this reviewing process there is an implicit comparison with industry-standard solid fuelled technology; the outcomes reflect favourably on MSR technology.

6 Challenges

Risks such as proliferation and failures of components are seen as challenges. Over-emphasis on corrosion issues is perceptible, arising from inadequate familiarity with continuing achievements of molten salt chemists and the past achievements, in particular at the Oak Ridge National Laboratory, since the late 1950s. Sources of funding for implementation and development of innovative nuclear projects in the UK present a major risk and some possible funding opportunities are reviewed.

APPENDICES

The following appendices are not available in the public domain. Please contact the authors for further information.

MSR Benefits Explained

An implicit comparison with industry standard PWRs is made

Safe	No meltdowns possible No large release of radioactive gases Reactivity reduces in the event of overheating Low working pressure
Less waste	Low amounts of waste created Radioactive for 100s of years, not 100,000s Fission products are removed online
Efficient	Reactivity increases as heat is removed - load following High efficiencies enabled with high temperatures High fuel burn-up
Fuelled & cooled with Liquid Salt	Good heat transfer properties Compact (installation below ground proposed) Scalable from small to large reactors
Fuel Cycle Flexibility	Can 'burn' both waste and weapon stockpiles Thorium as a fuel source for millennia possible Relatively low proliferation risk possible
High Temperature Heat	High thermodynamic efficiency Suitable for additional industrial uses (cement, desalination, district heating)

- These features contribute to an affordable source of clean low carbon energy
- Costing estimates indicate that plant capital costs can be on par with fossil fuels
- The concept has been demonstrated, proposals are ready to be developed today

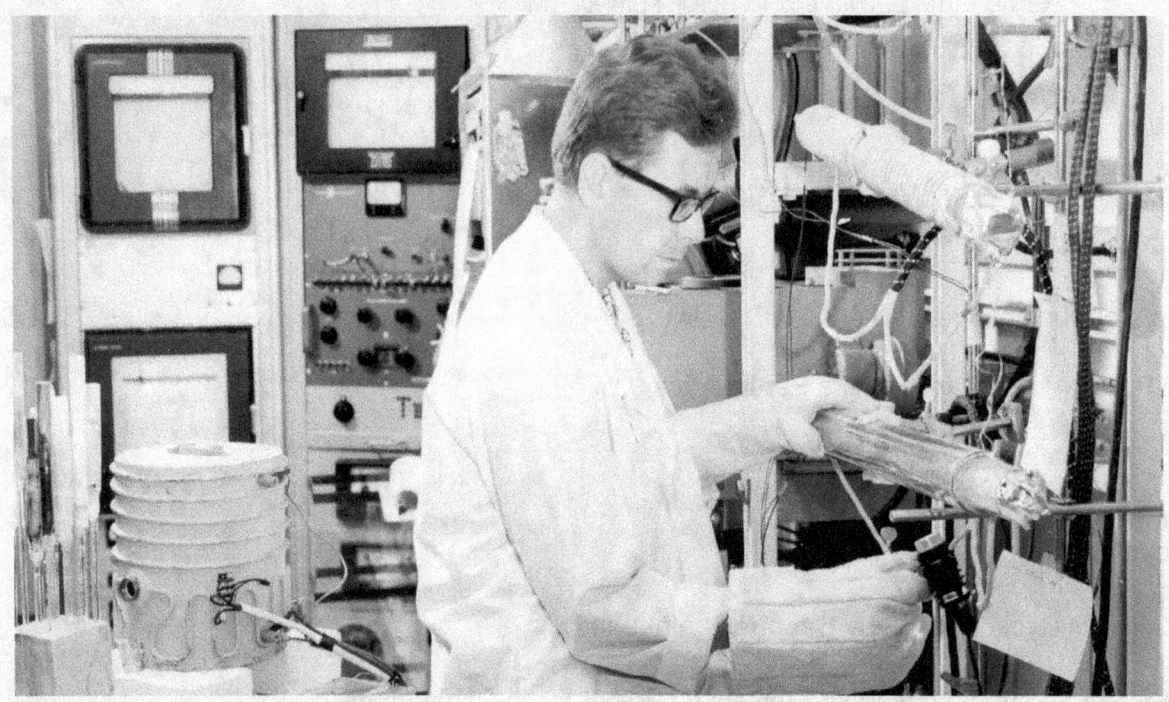

Dr. Trevor Griffiths, one of three directors of Energy Process Developments Ltd, heating a molten salt in a quartz tube at Oak Ridge as part of the MSR Experiment in 1968.

Introduction

Obtaining affordable energy for domestic and industrial use is a key activity in which the role of nuclear *fission*▲ is important. Engineers and scientists addressing this activity are becoming aware, particularly since global warming has become a concern, that the current nuclear technology poses serious difficulties in respect of affordability. The search for a way forward has created what amounts to a small international community of liquid-fuelled *molten salt*▲ reactor inventors and entrepreneurs. A central objective of the present study is to assess the technical, industrial, and economic opportunities provided by the individual commercial and institutional adherents to this community. These adherents, characteristically, are small to medium operators, not the established nuclear installation providers.

Energy Process Developments Ltd was initially created in response to a realisation that although there were compelling conjectures about the benefits of liquid-fuelled fission reactors, no-one anywhere was visibly planning to take the concept out of research and into implementation. These innovative devices, with unparalleled passive safety operation and the potential for reduced costs, were widely considered to be several decades away. The exception that emerged was a Chinese announcement in 2011 giving first priority to a pilot plant operating by 2015 – now postponed until after 2020. The outcome - seen as necessary because of lack of involvement elsewhere - was an application for government funding for this feasibility study.

Starting nearly a year ago, as part of an assessment of MSR activity internationally, members of the liquid-fuelled reactor community were approached. Proposals were received for pilot-scale implementation, where technical readiness was claimed. Six such specific proposals have been assessed by members of our study team and with commissioned expertise from established UK nuclear engineering firms. These proposals are seen as credible for the circumstances in the UK. One of these has emerged as most suitable for UK implementation.

The contents of this study report include a comparison of other advanced reactor concepts, a review of the historical and current background, information about liquid-fuelled reactors and the related science and engineering. The study activities included attending relevant meetings both in the UK and abroad to make presentations, and to meet academics, engineers and decision-makers. The project manager visited the Chinese researchers in Shanghai. A three man team - two directors and a very recently retired *Office of Nuclear Regulation*▲ safety inspector - travelled from Ontario to Atlanta to meet with molten salt reactor designers, with experts at the Massachusetts Institute of Technology, and also for a rewarding visit to the Advanced Reactor Systems and Safety Group at the Oak Ridge National Laboratory.

Significant components of this report comprise assessments by Frazer-Nash engineers of the six reactor configurations considered; an Ipsos-MORI online poll was conducted; Atkins provided economic study input; Caspus Engineering's review included fuel cycle proposals; a recently retired nuclear safety inspector, David Glazbrook, reported on the regulatory regime and availability of sites; Endorphin Software devised the procedure for recording and analysing expert opinion on the various proposals; and an insurance broker's opinion. All these contributions, with the exception of the survey, provided some or all of the work involved on a voluntary basis. A major part of the funding for this study came from the UK's innovation agency, Innovate UK. The study was based on information from the six designers and from many more. Sincere gratitude is due for all the help with these essential inputs and the support from the fledgling MSR community.

A glossary of terms exists at the end of this document which explains technical terms and concepts where the symbol ▲ is seen.

1 Opportunities & Industry Overview

Background

The UK has stepped down from a leading role in civilian nuclear technology that it occupied from the 1950s to the end of the 1970s. Two graphs summarise the period of decline:

UK Nuclear R&D Workforce: Showing the reduction in workforce following the closure of Government nuclear laboratories

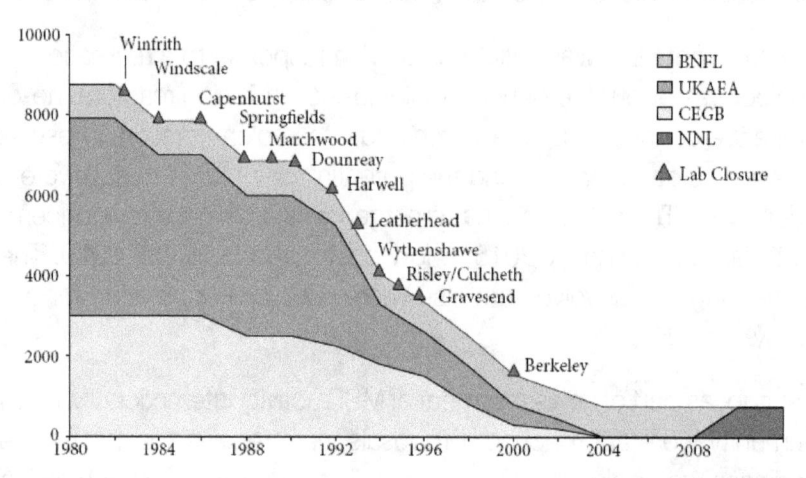

Reproduced from the Lords Science & Technology Committee's report

Civil nuclear R&D spend by country, 1980-2009

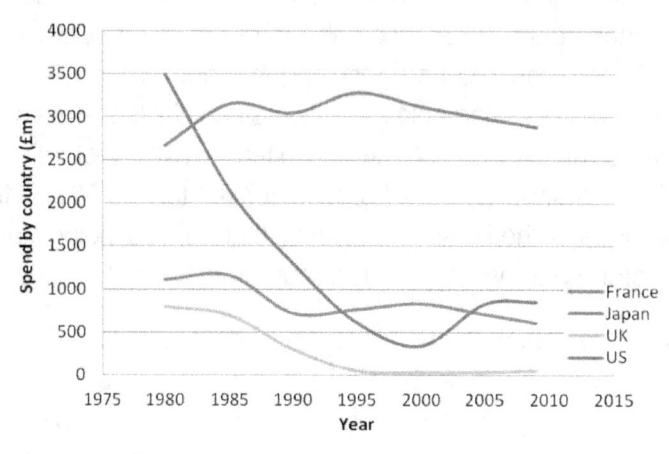

From HMG (March 2013), A Review of the Civil Nuclear R&D Landscape in the UK

The mismatch, embodied in this decline, between perceived global opportunities and UK science and engineering capability is dramatic.

In terms of UK capability, a government review in March 2013 reported there were 1,890 nuclear R&D personnel, mainly at the National Nuclear Laboratory, the Culham Centre for Fusion Energy, and the other still remaining National Laboratories. Of this total, just thirty-three people focussed on advanced fission reactor systems. In the universities there were less than 5 PhD students and just 0.2 full-time equivalent postdoctoral research assistants engaged on next-generation nuclear fission. Just under half of resources, in terms of both people and funding, are directed to *nuclear fusion*▲. Total, mainly government, funding in 2010/2011 on R&D for nuclear fission amounted to just over £30 million, with somewhat more for fusion R&D.

This nuclear fission capability was the UK's inadequate response to a world market estimated by 2030 at nearly a trillion pounds (WNA 2013). Of this total the UK could expect, if correct decisions were made in a timely fashion, the opportunity for up to a £240 billion share of this market. An alternative, if nothing much is done, is that the UK becomes a passive receiver of technology. Now today at the outset of a five year term of office, the UK Government has a unique and marvellous opportunity to initiate a new civilian nuclear era that will essentially be characterised by innovative reactor technology.

Current policy

Coincident with Sir John Beddington's retirement as Chief Scientific Adviser, the Government's strategy as a response to a 2011 House of Lords report was presented in March 2013. Six key policy papers by the Department for Business Innovation & Skills with contributions from the Department of Energy and Climate Change were published simultaneously on 26th March 2013 followed by miscellaneous reviews and reports[1,2,3,4,5,6].

A complacent theme repeated in these policy papers is that UK capabilities for nuclear engineering are world class. A scenario emerges of UK investment to provide 5 sites, starting with Hinkley Point C, with about 3 - 4 GW each. This amounts to a total 16 GWe new nuclear build by 2030. There is an expressed intention to follow this up by building up the nuclear power sector to 75 GWe by the middle of the century. Investment will be industry-led, with any Government contribution not clearly specified. A likely outcome, however, is for industry to choose more-of-the-same technology, that is, for solid-fuelled light water reactors. Difficulties are evident.

Difficulties

First is the issue of human resources. A recent skills report (HMG 2015) says the "national nuclear workforce is ageing and attrition rates are high". Industry's own research forecasts that the workforce must grow by 4,700 people a year over the next 6 years. Over the same period 3,900 people a year are expected to leave the sector, mostly because of retirement. Therefore the sector must recruit 8,600 people every year. In addition, more expert staff will be needed. Experts may need up to 20 years of preparation for some key posts. Another particular resources challenge that emerges according to recent information (mid-February 2015) concerns the Office of Nuclear Regulation. Currently it employs 306 inspectors, 254 of whom are safety inspectors. They are busy people. The Hinkley Point C Generic Design Assessment required £33 million for Office for Nuclear Regulation charges for 50,000 days of regulatory effort (together with perhaps twice that cost for the licensee)[7]. This represents a possible work requirement for some of the regulatory procedure for a single nuclear power plant. Another four such assignments are expected within the next few years needing a considerable increase in the number of inspectors.

Secondly, more-of-the-same new civilian nuclear build raises a challenging issue concerning long-term policy. Are these old-style reactor options fit for purpose? Over the first 60 years of the first nuclear era the inherently unsafe character of solid fuelled water cooled reactors has been addressed by increasing construction costs, both in time and money, to meet rising safety requirements. They are now the safest technology in the energy sector but the outcome is electricity generated at prices greater than with fossil fuels and widely perceived as unaffordable.

▲ *see Glossary*

Opportunities

The listing of the six Gen IV options gives an indication of a range of possibilities that provide options for the next reactor generation. Liquid-fuelled reactors – the subject of this study – are seen to provide more benefits than others (see Section 5, MSR Benefits for further discussion on this). With immediate and decisive commitment tangible achievement in a five-year time-span could be realised.

As this study reveals there are several small scale developers for liquid-fuelled molten salt reactors. Six of these have undergone technical review by the study team. Unlike the Gen IV liquid-fuelled reactor selected as the reference design – the Euratom molten salt fast reactor project based at Grenoble - each of the reactors included in this study are technically ready to take the next step towards industrial implementation. This immediate next step is to complete engineering design work for a prototype or pilot-scale reactor. What stops this happening right now is the high cost of any radioactive device.

Barriers to entry of new technologies have become extremely high as an unintended consequence of the complex regulatory system that has been made necessary by the need to assure the safety of intrinsically hazardous nuclear reactors, such as the Pressurised Water Reactor. This high barrier favours the incumbent technologies and creates a technology "lock in" situation.

Speculative investment involved in commissioning a single new MSR reactor project is reckoned in the hundreds of millions of pounds. A realistic assessment indicates this happening when and if Government makes a robust commitment to a significant role in providing funding. The probable reward for UK industry and the public is very large, and the cost to the taxpayer affordable.

Elsewhere in this report the Moltex Stable Salt Reactor is selected as providing the basis for a UK programme for development. A likely sequence of events is provided below, the time-line being dependent on financial and regulatory commitment. The evaluation in this report is comparable with the evaluation of other MSR proposals reviewed.

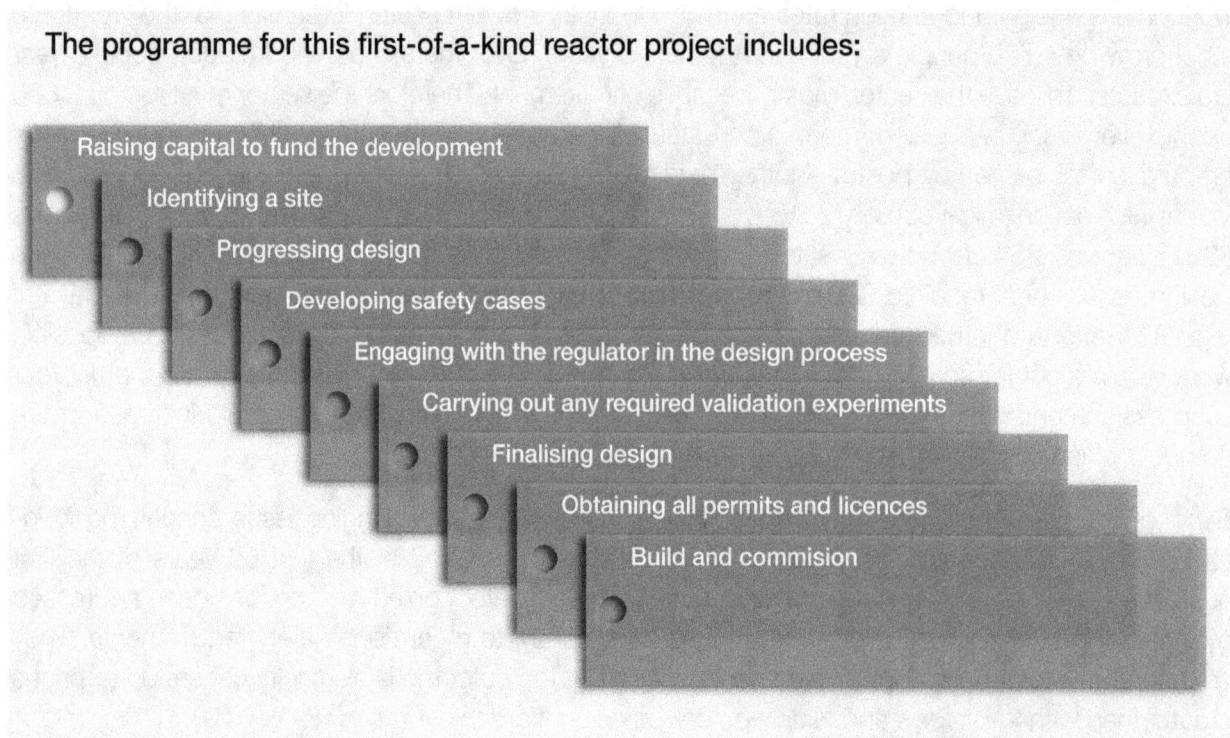

The programme for this first-of-a-kind reactor project includes:

Raising capital to fund the development

Identifying a site

Progressing design

Developing safety cases

Engaging with the regulator in the design process

Carrying out any required validation experiments

Finalising design

Obtaining all permits and licences

Build and commision

Benefits for private and government investment

An Atkins report commissioned for this study unsurprisingly stated that the case for supporting technology and construction of a pilot plant will only be of interest to investors if there is a long term financial benefit and a return on the initial investment.

The potential market for nuclear power is large enough to justify supporting development of next generation nuclear technologies providing benefits are sufficient. Molten Salt Reactor technology offers a number of benefits to make it attractive as a low carbon source of energy to investors compared with other nuclear technologies:

- Lower capital cost than other nuclear technologies
- Potential for IP ownership
- Lower levelised cost of electricity compared with other low carbon sources
- Passive safety features making it more acceptable to the public
- Co-generation opportunities that provide increased revenue
- A manageable nuclear waste legacy
- Scaleable to suit both the SMR market and the major installations

Small modular versions of the Molten Salt Reactor have additional benefits in that they can be manufactured offsite and brought on line more quickly. Funding can be phased with revenue realised earlier than with larger reactors, thereby reducing risk to investors and improving investor cash-flow. Smaller versions are more suitable for remote locations and for replacement of closing coal fired power stations. This would bring cost savings as existing grid infrastructure and potentially turbine halls could be upgraded and reused.

From a government perspective, investing in the development of Molten Salt Reactor technology in the UK will bring additional benefits to the UK economy including:

- UK leadership in Molten Salt Reactor technology
- Manufacturing growth
- Employment (engineering, manufacturing, construction, operations and maintenance)
- Contribution to GDP
- Increased revenue from taxes
- Less nuclear waste for long term storage
- Technology export potential
- Contribution to carbon reduction targets
- Burning plutonium to reduce the stockpile
- Burning other actinide wastes

The UK will expand its nuclear workforce recruiting graduates and apprentices into the sector. Early involvement in advanced nuclear technologies will progress the UK's capabilities and reputation as an international leader in nuclear engineering as per the HMG paper *The UK's Nuclear Future*, to be "at the top table of nuclear nations" and "a partner of choice for Gen III+, Gen IV and SMR technologies worldwide".

Prototype Reactor Uses

Throughout the lifetime of this initial prototype there are various uses foreseen:

- As a demonstrator, new Molten Salt Reactor technologies would be proven. This brings investor and public confidence in liquid-fuelled reactor implementation. Reactor design would allow it to transition to other operating formats - in the *thermal spectrum*[▲], as a thorium breeder or as a Low Enriched Uranium burner.

- As a test reactor, its function would include testing its own reactor components and mechanisms under real conditions.

- As a research facility, it would constitute a suitable neutron source for academic researchers to irradiate materials.

This prototype reactor could be used for research and development for as long as required over a range of operating conditions. However, after a relatively short period of operation as a demonstration reactor, it would be the basis for the next step, a second reactor for industrial application. Any prototype would be used as a training facility for operators of future, larger facilities.

The business outlook after the second reactor is operational is that investor payoff in terms of possible cash-flow commences. In the longer term more significant benefit is a real opportunity to establish what would become one of the industry-standard nuclear fission technologies.

Climate change and related global aspirations

There is a robust business case for committed UK involvement in the best proposed civilian nuclear fission technology, namely the Molten Salt Reactor. In addition, an underlying driver for effective and urgent action is a clear link to concern about climate change. To address the climate change issue sustainable and affordable carbon-free devices for harnessing energy are required for the UK.

When considering all nations in the world, personal poverty, energy poverty, world population growth and greenhouse gas concentrations are all apparently linked. Consider three sets of facts:

- Taking gross domestic product per capita of US$7500/y as the boundary between poverty and prosperity, a clear majority of nations in prosperity use on average, more than 2000 kWH/y per person (CIA Factbook)

- Nations in prosperity have fertility rates just below a stable population rate of 2.3 children per woman (CIA Factbook)

- Greenhouse gas concentration and world population have grown in parallel, close to exponentially, over the last 2000 years (International Panel on Climate Change and Wikipedia)

This evidence leads to a strong conjecture that a necessary condition for addressing greenhouse gas concentration reduction is world-wide access to affordable, sustainable energy. For this objective the very best technology must be implemented to provide energy for civilian use. It is often proposed that conservation of energy – changing light bulbs etc – would resolve the issue. However, if practised in just the wealthy fifth of the world, these measures – although they may be good in themselves and even cost effective - cannot bring a favourable global outcome when set against the energy requirements for raising the remaining four-fifths of the world out of poverty.

▲ *see Glossary*

Generically, specific characteristics of Molten Salt Reactors indicate that they are necessary for a successful outcome in addressing climate change as they can be engineered to be relatively affordable. They can be sufficiently compact to be secure, virtually self-sufficient requiring very little external resources including cooling water, and they are passively safe and load-following.

SECTION OVERVIEW

- UK has a non-existent nuclear R&D spend compared with other large countries.

- Opportunity exists for UK to have a £240bn share of £1tr international nuclear market by 2030. Liquid-fuelled MSRs can be developed in the UK to avail of this market.

- MSRs can have potential to be more affordable and are safer than today's technologies. MSRs can deal with the plutonium and waste stockpiles.

- Wealth, energy consumption, greenhouse gas and population growth are all apparently linked. With a source of cheap, clean energy, all can be stabilised.

- Immediate government action can enable MSR technology.

2 MSR Designs Assessed In This Study

A review of MSR activity internationally was carried out – see Appendix A. Following this, six *molten salt*▲ reactor designs were assessed for their suitability as a prototype demonstration reactor in the UK. The business plan and commercial viability of the vendors' proposals was beyond the scope of this study. All six have been evaluated and are deemed fit for purpose. They are each at an early concept stage, ready to be developed further towards detailed design and experimental validation. All designers have been met and were consulted throughout the review process. Some proposals require more experimental work than others to validate innovative solutions.

These proposals were selected on certain criteria which include:

- having a reactor design with a molten salt as fuel and coolant
- having a commercially active team prepared to implement or license their development of a pilot MSR in the UK
- being technically ready to design and license
- designs that could have a pilot-scale version on line in less than ten years

Technical appraisal of the designs was carried out by Frazer-Nash Consultancy, Caspus Engineering and Energy Process Developments Ltd including voluntary involvement by each entity. Teams from other international institutions familiar with molten salt reactors assisted with the final evaluation by Energy Process Developments.

An example of the assessment criteria included:

- the availability of materials, fuels and salts
- safety characteristics
- ability to treat spent nuclear fuel
- development programme
- suitability as a demonstrator
- proliferation risk
- external and internal hazards
- environmental impact and sustainability
- licensability
- UK Intellectual Property development potential

The full evaluation is in the appendices with restricted circulation for confidentiality reasons. Please contact Energy Process Developments directly for further information. Below is a brief summary of each design in alphabetical order, a longer summary is available in Appendix A with a table comparing the main characteristics of each design.

▲ *see Glossary*

MSR 1 – Flibe Energy – Liquid Fluoride Thorium Reactor (LFTR)

Flibe Energy, one of the first to resurrect the molten salt reactor concept, and based in the USA, proposes a 2MWth two fluid breeder design. It is based on work carried out by the Oak Ridge National Laboratory team in the 1970's. It operates in the thermal spectrum moderated by graphite. Its fissile element is uranium-233 which is bred from thorium in a blanket salt at the outer edge of the reactor core.

As with several other proposals, it is planned for the reactor and drain tanks to be located underground. Being a breeder reactor the LFTR design utilises the full potential of thorium. It is a valid proposal which can be taken forward for development. Little experience exists with the specific reprocessing methods required.

More information: www.flibe-energy.com

MSR 2 – Martingale Inc. – ThorCon

The ThorCon design is a single fluid thorium converter reactor that operates in the thermal spectrum. It is in principal similar to the MSR Experiment and its fuel is *denatured*[▲] using a combination of U-233 from thorium and U-235 enriched from mined uranium. Its core is graphite moderated and the full scale version runs at 550MWth. A centralised facility is proposed to reprocess the spent fuel salt from multiple plants. The design team comes from a shipping background and brought in nuclear expertise from members of the MSR community from across the United States. As a concept, a pilot-scale version of this plant would be similar to the MiniFuji, a concept that the Japanese have been working on for a long time. Both designs are replicas of the MSR Experiment with minor alterations and no new technology. The review carried out for this study assessed both concepts primarily utilising the more available ThorCon data.

One major benefit for this design is that the concept is proven. This brings the possibility to go straight to a larger plant if the experience from the MSR Experiment is used. The designer's proposal is to prefabricate 500 tonne components that get barged to the power plant site. This has the potential to dramatically reduce costs and improves the quality.

More information: www.thorconpower.com

MSR 3 – Moltex Energy – Stable Salt Reactor (SSR)

The Stable Salt Reactor has a design team based in the UK. It is a fast spectrum pool type reactor. Unlike all the other design concepts considered, its fuel is static and is not derived from the two molten salt reactors developed at Oak Ridge National Laboratory. Its static fuel concept was actually correctly rejected as unsuitable for an aircraft borne reactor by ORNL and apparently never reconsidered when the program moved to ground based reactors. The full size version is proposed at 1GWe and the prototype at 150MWth, but run at a lower power.

The fuel salt is derived from spent nuclear fuel in a sodium chloride solution that sits in tubes in a pool of coolant salt. These are arranged in arrays similar to conventional solid fuel arrays in a light water reactor.

▲ *see Glossary*   *MWe = electrical power   MWth = thermal power*

Being a static fuel, there is no need for pumps or other devices to control the flow. Fast reactors are more suited to burning the long lived troublesome actinides in spent nuclear fuel. Complications with the use of graphite are also avoided. Experimental data of the heat transfer properties of the fuel salt tube to coolant will be required to validate computational fluid dynamics simulations.

For the reasons above, and particularly its relative simplicity and relatively few and low technical hurdles, this proposal is deemed to be the most suitable configuration for immediate pilot-scale development in the UK.

More information: www.moltexenergy.com

MSR 4 – Seaborg Technologies – Seaborg Waste Burner (SWaB)

The SWaB prototype proposal is a 50 MWth single fluid reactor that operates in the thermal-epithermal spectrum. It is graphite moderated and fuelled by a combination of spent nuclear fuel and thorium. The design team based in Denmark, is a combination of physicists and chemists from the Niels Bohr Institute and the Technical University of Denmark. It is designed to take spent fuel pellets directly for de-cladding and insertion into the fuel salt.

This design has innovative features and is suitable for modularised construction. It is in its early stages of development but has a competent and motivated design team. The reprocessing system brings many benefits but would delay the deployment of a first-of-a-kind pilot MSR. This is a suitable device for efficient utilisation of the spent nuclear fuel stockpiles. The concept is appropriate for further development.

More information: www.seaborg.co

MSR 5 – Terrestrial Energy – Integrated MSR (IMSR)

The Integral MSR is also based on the MSR Experiment but has been modified to have a more sealed, passive approach. The design team is based in Canada with international involvement and support. An 80 MWth prototype reactor is proposed.

Operating in the thermal spectrum with a graphite moderator inside the sealed unit, it can fit on the back of an articulated truck. This unit contains the fuel salt, moderator, heat exchangers and pumps. The plant is fuelled with 5% low enriched uranium where the U-235 is *denatured*▲ with U-238. This core is modular, designed for a high power density and replacement after a seven year cycle in a plant with an overall lifetime of over thirty years.

This 'seal and swap' approach reduces on site complications and risks. Using low enriched uranium, it has a fuel cycle with which regulators are familiar. This proposal is suitable for developing fully and launching commercially immediately.

More information: www.terrestrialenergy.com

MSR 6 – Transatomic Power Reactor (TAP)

Transatomic Power's proposed design is a 20MWth demonstration reactor which is similar to the MSR Experiment except for its utilisation of zirconium hydride (instead of graphite) as a moderator and LiF-based salt (instead of a FLiBe-based salt). These changes enable a twenty fold increase in power density and the use of very low enriched fuel. It operates in the thermal spectrum and with a

▲ *see Glossary*

significant neutron flux in the fast spectrum. It is a single fluid configuration. The design is from a start-up company based in the US that originated out of Massachusetts Institute of Technology and still has strong links there.

The commercial version of the TAP reactor is 520MWe and can be fuelled by low enriched uranium or spent nuclear fuel, thereby reducing the amount of long-lived nuclear waste by converting it to much shorter-lived fission products.

Experimental work is progressing at the moment on the use of zirconium hydride as a moderator and its cladding materials. The application of novel materials brings benefits, particularly its efficient utilisation of spent nuclear fuel but it also sets it at a longer time to deployment than more traditional configurations. This design is innovative and suitable for full development.

More information: www.transatomicpower.com

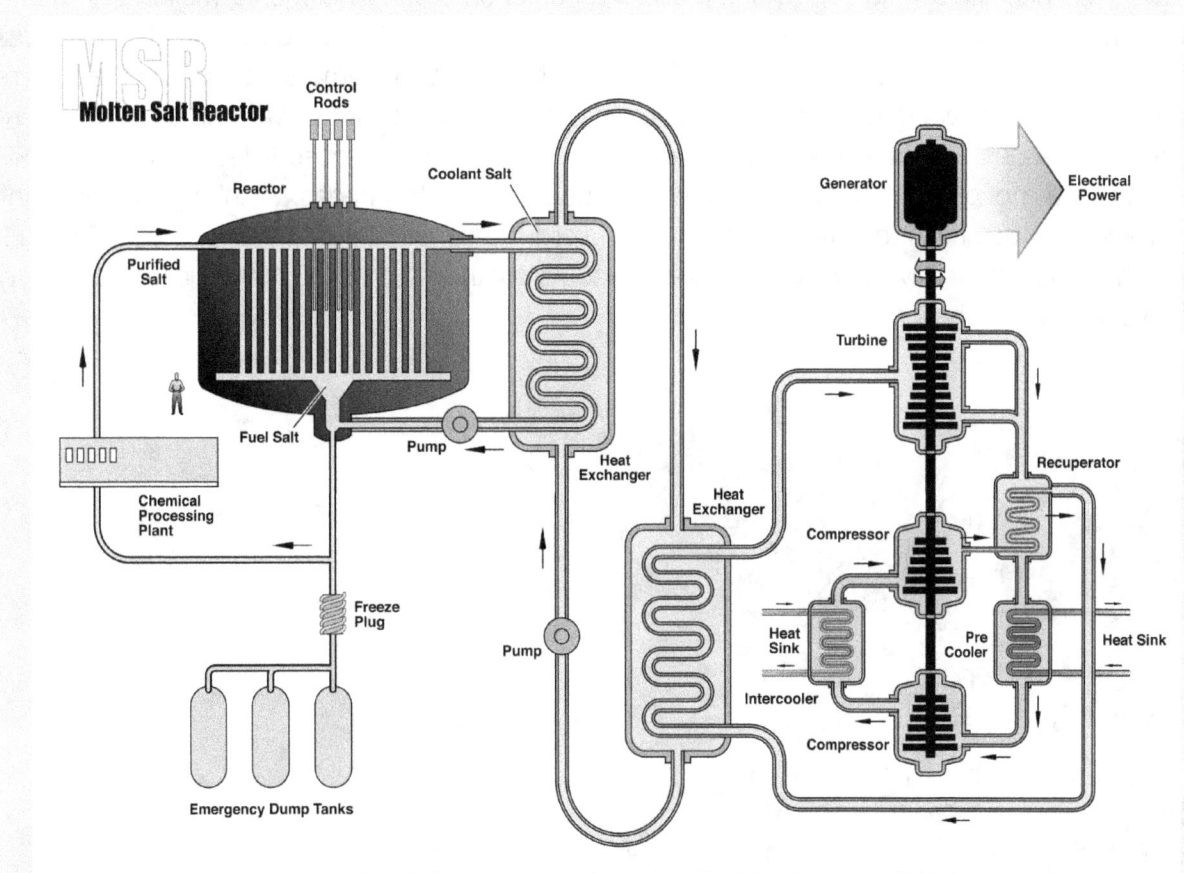

Schematic for a generic MSR showing commonly applied features. Image reproduced from www.gen-4.org

SECTION OVERVIEW

- Global MSR research and development is currently led by China. Elsewhere, small start-ups with innovative new designs show promise.
- Six proposals reviewed for suitability as demonstration prototype in the UK. All are seen as valid proposals at this early stage of design.
- The Moltex Stable Salt Reactor brings simplicity and advantages for the UK in particular.

3 Historical Background

The birthplace of the Molten Salt Reactor was the Oak Ridge National Laboratory, Tennessee, (built as part of the wartime Manhattan Project). Dr. Alvin Weinberg was first its main designer and then laboratory director. He designed other reactors that were water-cooled but he preferred the MSR because of its safety features and because it did not produce plutonium.

Dr. Weinberg, as director at Oak Ridge National Laboratory, was able to develop liquid-fuelled reactor technology, having brought together some 300 top scientists, many of whom were chemists. The first MSR operated for about 100 hours. It was an experimental trial for an idea that nuclear energy could power aircraft to deliver nuclear weapons. Weight restrictions indicated the choice of liquid-fuelled reactors operating at atmospheric pressure. The project was abandoned because radiation protection for the crew required heavy lead shielding, liquid fuel, and coolant handling was difficult in flight, and crucially, Wernher von Braun's ballistic missiles were preferred to long-range bomber aircraft.

The aircraft experimental reactor provided the forerunner for an 8MW experimental molten-salt-fuelled reactor. The MSR Experiment was constructed at Oak Ridge by 1964, went critical in 1965, and was operated until 1969, when the project was ended abruptly by the U.S. administration. Alvin Weinberg's tenure of the post of laboratory director came to an end, partly because he was expressing concern about safety issues associated with solid fuelled reactors. Further proposed development of MSR technology was not pursued as it offered nothing for perceived military needs. Weinberg, among many achievements, was one of the principal inventors of solid-fuelled pressurized water reactors (PWR). It was a PWR that was ready when needed as a power unit for the Nautilus, a submarine with capability to deliver nuclear warheads.

The Aircraft Reactor Experiment reproduced from the Oak Ridge National Laboratory newspaper. The reactor core is shown rising out of its housing to assess the radiation levels emitted at varying distances away.

The Nautilus, commissioned in 1951 and launched in 1954, became the successful prototype for subsequent nuclear powered submarines. Its pressurized water reactor system was subsequently constructed on land as the Shippingport Reactor to provide electricity for public supply, which initiated the development of the 440 civilian nuclear solid-fuelled reactors in use today.

Molten salt reactor technology fell victim to perceived cold war military requirements, which included plutonium production. US government policy decisions shifted MSR research and development away from ORNL to California, where they were allowed to die. The MSR became an unattractive disruptive alternative to solid-fuelled technology with expected adverse effects on established nuclear industries.

Some ten years before his death, in 2006, Alvin Weinberg in a talk to Korean scientists[8] recalled that regular meetings, known as 'New Piles Seminars', took place at the Chicago Metallurgical Laboratory in 1945, just after the design of Eugene Wigner's Hanford Pile was finished, and after Fermi had established the first chain reaction. A particular point of interest was whether chain reactors that might have civilian application be considered as chemical or mechanical engineering devices. Both Wigner and Urey insisted they should be regarded as chemical devices, specifically devices in which the fuel elements are replaced by liquids. Wigner, along with others present, identified the advantages of molten fluoride salts.

United Kingdom government supported funding for the vigorous activities that paralleled US nuclear research up to the late 1960s. UK funding for civilian nuclear power research, development, and deployment, was not evident before that. This early period saw the UK at the cutting edge for gas-cooled reactors and similar innovations. What actually continued was secret or confidential work at several sites such as Risley, linked to the UK Atomic Weapons Research Establishment. By 1976 this work had been shelved and has since remained undisturbed in the National Archives. Now declassified, it is currently under review. A notable achievement from the 1970s was a well-developed design for a large (2.5GWe) helium-cooled fast spectrum MSR with the fuel dissolved in a molten chloride salt. Lead was also reviewed as a coolant. It is only now that these designs for this much needed innovative technology are again under scrutiny, both here in the UK, and internationally.

SECTION OVERVIEW

- The Pressurised Water Reactor was developed for the defence program and continued on to produce electricity.
- MSR as a concept was successfully demonstrated in the 1960s. It did not meet defence requirements and was stopped.

4 An Introduction to Liquid-Fuelled MSR Technology

Nuclear power is harnessed from nuclear chain reactions where a captured neutron *fissions*▲ a fuel nucleus, releasing new high energy neutrons to sustain the reaction. Fission of a uranium atom releases a hundred million times more energy than the combustion of a carbon atom. Fission is induced only when a nucleus is bombarded with neutrons, gamma rays, or other radiation. Fission produces atoms of lower atomic mass. These fission products are usually radioactive. The mass of the fissioned atom is greater than the mass of the fission product atoms. This difference in mass is converted into energy according to Einstein's equation $E=mc^2$ where E is energy, m is mass and c is the speed of light. This energy provides the heat to ultimately produce electricity.

Slow moving neutrons have a larger probability of inducing fission than fast ones. Many reactor configurations slow down the emitted high energy neutrons to ensure efficient capture to sustain the reaction. An effective moderator used for this purpose is graphite as it can be placed in contact with the fuel, whether solid or liquid. Other moderators, including water, need protective cladding.

Most molten salt-fuelled reactor designs operate with slow, that is *thermal spectrum*▲, neutrons. They have employed blocks of graphite as moderator through which the fuel is pumped. Neutron economy requires that any chemical that absorbs neutrons should be avoided both in construction and in the fuel supply. This presents many challenges. One such challenge is procurement of graphite of a nuclear grade. Recent surveys identify a shortage of supply of the high quality special coke produced by oil companies which is needed as starting material for suitable graphite manufacture. Because demand for this is very small there is at this time insufficient interest in providing a supply.

Materials for many reactor components and design materials utilize the characteristics of hafnium and zirconium, which occur naturally together and are physically very similar. However hafnium absorbs 600 times more neutrons than zirconium and so is used in reactor control rods. Pure zirconium is almost transparent to neutrons and hence is desirable as a construction material for many reactor components (this has included metallic cladding of fuel rods for solid fuelled PWRs).

A thorium/uranium fuel cycle in an MSR, when configured appropriately, can breed more fissile fuel than it consumes. This class of reactor is known as a breeder reactor (see *Thorium*▲ for further explanation). This operation requires reprocessing of the fuel salt. Less moderated reactors increase actinide burn-up in the fuel salt. Un-moderated fast reactors work well with plutonium fuel. A fast, high neutron energy spectrum reactor is effective at burning long lived actinides. Using thorium in the reactor fuel produces less long lived actinide waste than the uranium/plutonium cycle. These reactor characteristics are considered in the review process of this study. Most designs can be re-configured to accommodate different fuel cycles. Further advantages are discussed in the next section.

Except at start up and shutdown molten salt reactors operate in a near equilibrium steady state. With negative reactivity coefficients they slow down or speed up to follow load. Fuel quantity is adjusted by breeding or replenishing. Refuelling is often planned as part of online reprocessing.

Molten salt reactors employ several methods of fission product removal. Noble gases are removed either by bubbling out independently or assisted with a helium sparge. Fluorination - bubbling fluorine gas through a fluoride salt - removes various elements. This procedure was studied extensively at ORNL[9]. Separated elements and compressed gases are stored to decay until safe. Some metals plate-out on vessel boundaries or on nickel mesh filters in a side stream. Lanthanides are removed by a variety of methods, an example being reductive extraction using liquid metal containing a reducing element and a *molten salt*▲ containing an element to be extracted.

▲ *see Glossary*

A status update of various pyro-chemical processes was published by OECD in 2004[10] and although valid today there are subsequent improvements in several areas, e.g., Griffiths and Volkovich, *Nuclear Technology*, 2008, **163**, (3), 382-400

Molten salt has a viscosity similar to water and is transparent.

SECTION OVERVIEW

- Nuclear energy has a far higher energy density than other sources.
- MSRs have their fuel dissolved in a liquid salt which brings many advantages.
- They operate in the fast or thermal spectrum with a wide range of fuel cycles.

5 MSR Benefits

Public opinion, evaluated in a recent Ipsos MORI online survey for this project, indicates that the following is the priority with which different characteristics are ranked in the UK by those who approve of new nuclear build:

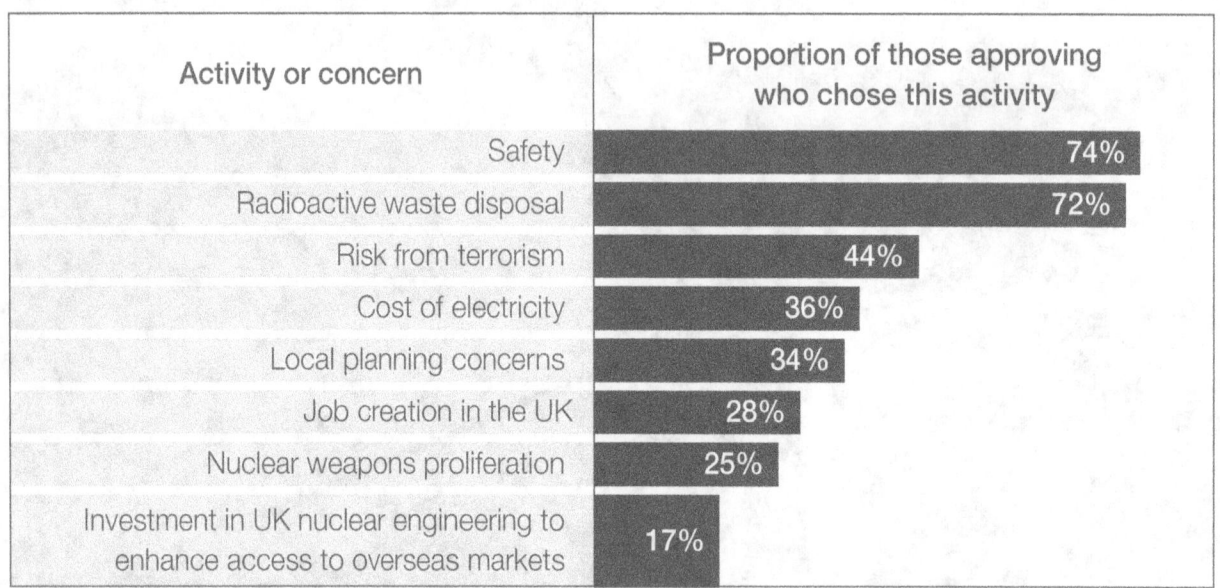

Activity or concern	Proportion of those approving who chose this activity
Safety	74%
Radioactive waste disposal	72%
Risk from terrorism	44%
Cost of electricity	36%
Local planning concerns	34%
Job creation in the UK	28%
Nuclear weapons proliferation	25%
Investment in UK nuclear engineering to enhance access to overseas markets	17%

In the following text the benefits are considered in the order above. Other benefits follow.

Safety

Generically liquid-fuelled *fission*▲ reactors are passively safe, principally on account of an inherent negative reactivity coefficient. This arises from reduced neutron capture for several reasons including thermal expansion as the fuel temperature rises. If the fuel temperature increases the reactivity decreases. This has an additional advantage in load following capability as the reactivity increases when heat is withdrawn via the heat exchangers. (More information is available[11].) These reactors work at near atmospheric pressure, eliminating the risk of explosion and large release of volatile radioactive substances. Melt-down events do not occur as both fuel and coolant are already in a liquid state well below boiling point. Typically there is provision to empty all the liquid fuel by gravity into tanks configured not to allow criticality either in an emergency or for routine shutdowns. A device such as a *freeze plug*▲ in the outlet allows this.

Radioactive waste disposal

With liquid fuel, very high burn-up is achievable, typically over 90%, leaving relatively small quantities of radioactive waste. This compares with solid fuel material which is rendered inactive after only a few percent burn-up, leaving a lot of radioactive material for disposal. High burn-up with liquids is possible because the fuel can be processed during operation, both by removing or adding components in order to maintain criticality. This can involve full chemical online reprocessing or lesser interventions. With liquids, troublesome inert gases can be removed continuously, an operation not achievable with solid fuels. The quality of the radioactive waste produced, particularly in respect of half-life, is very significant in the safe handling of nuclear waste. Each particular fuel element fissions to produce a characteristic family of actinides. Generally, the higher the atomic number of a fuel element, the longer are the half-lives of the remaining actinides. Thorium bred to uranium-233 on fission creates waste with half-lives of the order of 30 years. This means that radio-activity is effectively zero after about 300 years. To

▲ *see Glossary*

summarise: MSRs produce relatively small quantities of waste that should only require a few hundred years of safe management rather than thousands.

Risk from terrorism

All civilian nuclear power plants are now treated as being at risk. However, MSRs have only a small footprint compared with current industry-standard pressurised water reactors. This allows the reactor itself, or 'nuclear island', to be placed in a hole in the ground. This gives better protection against attacks from the air or from cunningly placed explosives. Furthermore, in an underground installation incident, any leaking container will be isolated. Another major consideration is that excess reactivity is not required with liquid fuels. Unpressurised, these reactors are not inevitably going to spread radioactive material into the atmosphere. The other recognized risk is proliferation of weaponisable radioactive material. In this respect the MSR is widely acknowledged to present relatively low risk. The fuels of choice can always be mixed with other isotopes or chemically similar material that would make enrichment a requirement for a weapon. Pure uranium-233, bred from thorium, is sometimes identified as a real danger. It has been tried as a weapon but is difficult to use – it is never seen as a material of choice for weapons.

Cost of electricity

In brief, electricity costs derive from amortised total investment costs, fuel cost, and running costs. Running costs are broadly similar for any energy harnessing procedure, fuel costs are lower by a factor of about 30 compared with solid fuelled reactors because of the difference in burn-up, and by current estimates, the investment cost for an MSR can be engineered to be lower than for solid-fuelled reactors. Electricity from some MSR plants can be competitive with fossil fuel plants. This arises principally from simpler engineering without the need for high integrity pressure domes and associated vessels.

- Cooled with molten salts
  - ↳ Low operating pressure
    - ↳ Not pressurized
      - ↳ Compact
        - ↳ Small footprint
          - ↳ Scaleable
            - ↳ Low investment cost
              - ↳ **Affordably low cost for output energy**

- Safe because of negative reactivity coefficient
  - ↳ Safe because drainable to alternative container where chain reaction ceases
    - ↳ Capable of installation of radioactive process equipment below ground level
      - ↳ Reduced cost for provision of additional components and containment to ensure safe operation
        - ↳ **Affordably low cost for output energy**

- High burn-up of fuel achievable
  - ↳ Very low amounts of process waste produced
    - ↳ **Affordably low cost for output energy**

- Flexible requirements for fuel cycle
  - ↳ Possibility for low proliferation risk
    - ↳ Legacy waste burning
      - ↳ Low fuel costs
        - ↳ **Affordably low cost for output energy**

- High temperature heat
  - ↳ High thermodynamic efficiency
    - ↳ Possibility of air-cooling or low requirement for cooling water
      - ↳ Alternative uses as process heat to make cement, synthesizing hydrocarbons, water desalination and district heating
        - ↳ **Affordably low cost for output energy**

- **Affordably low cost for output energy**
  - ↳ UK markets
    - ↳ Global markets
      - ↳ Tremendous business opportunities for the UK

Low electricity cost plus safety are the important benefit of the MSR, both for the commercial opportunity potential and addressing climate change.

Local planning concerns

All civilian nuclear installations can be expected to generate opposition, some expressing rational concern about the environment, some more directed to apprehension about poorly understood potential dangers. In terms of environmental impact, there are smaller inputs of radioactive fuel and less output of radioactive waste. This results in a lower total activity content per installed power. All other impacts could be expected to be similar to any comparable electricity generating installation. Local disruption during construction is minimized if modular components are constructed elsewhere for onsite assembly.

Job creation in the UK

Setting up a new industry-standard liquid-fuelled nuclear power industry would provide many job opportunities at every level. It would, however, diminish activity in current nuclear fuel fabrication, waste handling, and re-processing. Decommissioning, a very labour intensive operation, would almost certainly be maintained or increased for many years to come.

Nuclear weapons proliferation

Proliferation is a term applied to access - by the wrong people - to nuclear weapons, fissionable material, and weapons-applicable nuclear technology and information. The 'wrong people' include nations not signed up to the Nuclear Non-proliferation Treaty and to individuals or groups with malign intent. Innovative nuclear technology, and the MSR in particular, is hardly relevant to any of this agenda as it is not a useful route to production of weapons grade substances.

Investment in UK nuclear engineering to enhance access to overseas markets

An urgent UK commitment to implementation of industry-standard liquid-fuelled reactors would restore the UK to its world leader status in cutting edge civilian nuclear power engineering, as it was in the 1960s and 1970s. The foreseeable world market for harnessing sustainable and carbon free energy has been estimated at about one trillion US dollars. The UK can adopt MSR technology to secure a substantial share of this opportunity.

High temperature process heat

Carbon-free, safe, abundant, available, sustainable and affordable heat can be obtained from the MSR. It promises great benefits including, for example:

- Cement, such as Portland cement, is currently made by a process with a very high carbon footprint, partly from the chemical process employed but largely from the amount of process heat required. This is an industry set to benefit greatly from affordable carbon free energy.

- The high temperature of heat available will enable the replacement of fossil fuel products, including motor fuels with synthesized hydrocarbons using recognized industrial practice.

- Electricity can be co-generated with large scale heat uses such as for district heating and for water desalinisation.

Comparison against Other Gen IV Technologies

1. *Very high temperature reactor (VHTR)*

   This is based on TRISO fuel technology and as a result operates at low power density. The prismatic or pebble fuel is currently considered non-reprocessable and must therefore go to geological disposal. The volume of that fuel is about 30 times greater than that of PWR fuel for the same power production making the disposal challenge rather larger. The lower decay heat production per unit volume of the fuel compared with PWR fuel is an advantage that could allow closer spacing of the spent fuel in the repository. Since all plans seem to allow the fuel to decay to the point where heat production is quite low before transferring it to a repository this advantage may not be any help. DECC estimate nearly £500,000 per AGR fuel assembly for geological disposal so the cost of disposal of a 30 times greater volume would become a real economic factor. On a global scale, looking at half of electricity production from the VHTR, this disposal problem becomes very serious indeed.

2. *Sodium cooled fast reactor (SFR)*

   This concept is intrinsically complicated to engineer safely and prone to fires. Quite a few of these have been built and they have all cost far more than PWRs. It is unrealistic to expect that the cost problem is dramatically reduced, given the high hazard level that is being contained.

3. *Supercritical water cooled reactor*

   Experience with coal-fired supercritical systems is that the cost of electricity does not fall compared with subcritical steam systems. The reason is that the extra capital cost cancels out the fuel efficiency savings. For coal, this is worthwhile as it reduces carbon emissions but for nuclear with its lower fuel costs, the effect would be to make the electricity more expensive than from a non-supercritical system.

4. *Gas cooled fast reactor*

   This is essentially a sodium fast reactor with the sodium coolant replaced with helium. As such it is an improvement but the need for large high pressure systems would be likely to add just as much cost as was eliminated by removing sodium. The fundamental hazards are still in place with high levels of volatile dangerous fission products and high pressures that can disperse them if containment fails. Massive containment would therefore be needed making cost lower than PWRs unlikely.

5. Lead cooled fast reactor

This design is considered along with lead cooled graphite moderated thermal reactors. Both are quite good at a fundamental level and do have real promise, however the challenge is material properties. The lead or lead/bismuth coolant has proved to be incompatible with long reactor life (lots of Russian experience) and novel, non-metallic, materials will probably be needed to achieve that long life. Once that is done however, the system is potentially intrinsically safe and low cost. Unfortunately, most of the knowledge is locked up in Soviet era vaults making development harder and the need for new, non-metallic, materials would make development a long, difficult process.

SECTION OVERVIEW

- MSRs can be designed with full passive safety and no possibility for widespread dispersion of radioactive substances.
- They have high burn up of fuel and produce little long lived waste.
- Plant costs can be on a par with fossil fuels.
- More advantages are apparent compared with other existing and advanced configurations.

6 MSR Challenges

Although MSRs have many benefits, implementation involves risk for a first-of-a-kind development as any apparent failure may terminate any programme for further MSR activity. The first operational plant consequently will be brought on line with great caution.

Alongside any technical challenges, obstacles include an established solid fuel reactor industry complete with supply chain and regulatory regime. Reactor chemistry is one of the relatively unfamiliar skills that need to develop in an already overburdened setting.

The reason for selection of other Gen IV designs than the MSR is often cited as building on experience. However, with a successful MSR experiment in the 1960s, the concept has been demonstrated and a useful amount of experience was acquired. Fortunately, the concept of *molten salt*▲ reactors has been kept alive by researchers around the world. Together their research has revealed that little of the required process science and technology remains unknown.

Internationally, regulatory changes to deal with radically different design features are needed. In the UK, education and training within the ONR can be put in hand in a timely manner with tangible government support.

Obtaining funding is potentially the biggest challenge in the development of any new technology: this is no different for MSRs. This major topic is discussed later.

Chemical Reprocessing Plant & Proliferation Resistance

Any need to operate an on-site chemical plant to manage core mixture and remove *fission*▲ products presents a challenge. This procedure has not been developed commercially. Some existing MSR proposals do not use online reprocessing, thus enabling a faster roll out. Online removal of fission products with *thermal spectrum*▲ configurations allows a better neutron economy. On-site chemical plant difficulties and costs of transporting radioactive materials are much reduced or eliminated with online processing.

Proliferation resistance is decreased when the fissile U-233 is separated from other elements on a reprocessing loop. However, the intense gamma radiation emanating from the decay products of U-232, and the need to separate it from U-233 before the U-233 could be used in an explosive device, greatly reduces its potential for terrorist activities. Handling is significantly more complicated than plutonium and the decay products are difficult to conceal.

US Government bomb makers have not succeeded with U-233. From their trials U-233 has been ineffective as a component. The explosive energy achieved was 40% rather than an expected 95%.

The production of weapon-grade nuclear material from a breeder MSR would require support to a level equal to a large government. Other known routes, including uranium enrichment, are likely to be preferred. International aims today are to reduce nuclear armaments and "burn up" plutonium and other actinides. A good way to do this is using these actinides as fuel in MSRs. A scenario where terrorists take over a breeder MSR and have the skills, apparatus and time needed for a result is an unimaginable scenario.

▲ *see Glossary*

Corrosion & Materials

Corrosion is often stated as the reason for the original MSR Experiment being stopped. PWRs and BWRs both have active programmes to investigate and control water-induced corrosion problems. Corrosion of metals by molten salts (often termed Hot Corrosion), in coal-fired power stations for example, has been under investigation and is now under control: one technique uses co-extruded tubing. The molten salt is formed on superheater tubes that operate at around 660°C, the same temperature as MSRs.

At Oak Ridge National Laboratory, a group of scientists in the MRSE programme focussed on understanding hot corrosion by fluorides and developing alloys that would sufficiently resist such corrosion. Initially they used Hastelloy C from which they developed *Hastelloy N*▲, a high-temperature nickel alloy that withstands aging and embrittlement. It resists oxidation for continuous service up to 982°C and for intermittent service up to 1038°C. It can be readily fabricated and is ideal for high-heat applications and equipment used with hot fluoride salts. After some time however, it was discovered that its internal structure could be significantly damaged by exposure to neutrons. Sadly this fact is used by critics of MSRs to decry the Oak Ridge design but further research by the laboratory's scientists reported a treatment that would prevent this. Experience is needed on welding techniques for Hastelloy N, something in which the Chinese are actively engaged.

Today, we have a much better understanding of how to control the redox potential of the salt solution. Once this can be monitored, corrosion effects can be greatly reduced and standard stainless steels can be employed. Several universities internationally are currently studying this: further experimentation will be required for regulatory justification before a plant is built.

Generally the fuel salts selected are readily available. An exception to this is the widely employed lithium-7. Large scale enrichment processes are no longer available and will be expensive.

Liquid Fuel & Irradiated Components

When the fuel salt is fluid, more components are affected by neutron bombardment than in a solid fuelled reactor. Manufacturing pumps and valves capable of operating at high temperatures, intensely radioactive and corrosive molten salt for long periods is not an insurmountable challenge but is relatively new and will require thorough testing. Similarly, consideration will be required when creating mechanisms and protocols to inspect safely and replace these pumps and valves when they have become highly radioactive from exposure to the fuel salt and its associated neutron flux. IAEA techniques for measurement of fissile content in a liquid need to be developed.

An advantage of a mobile fuel is that it can be removed from the core, rapidly rendering it subcritical. Demonstration of this process using appropriate valves is required to prove high integrity and immediate response. Decay heat builds extremely quickly for high power density configurations and safety features must be designed accordingly.

Designers must ensure there are no 'cold spots' as salt freezing can cause detrimental effects on components and heat exchangers.

▲ *see Glossary*

Funding Challenges

Obtaining funding is potentially the biggest challenge in the development of any technology, and this is no different for MSRs. It becomes apparent that funding for large scale projects such as prototype reactor development does not exist. A gap between research funding and industrial implementation is evident. Opportunities do exist for part of the process such as early design studies. Private and commercial investment routes are not discussed. The opportunities for government or a private investor are discussed in Section 1. Below is a description of some possible routes for funding in the UK.

Innovate UK is the UK's innovation agency, an executive non-departmental public body, sponsored by the Department for Business, Innovation & Skills. It is competition based and intended to support new technologies and improvements to existing processes to enhance sustainable economic growth. It intends to fund projects that could not otherwise get off the ground. Higher percentages of project funding are available for smaller enterprises. The largest grant available in a single energy competition to date is five million pounds although larger amounts have been awarded in other sectors.

Horizon 2020 is the European Commission's program to fund technologies to achieve the 2020 goal of 20% reduced greenhouse gas emissions, a share of renewable energies at 20%, and a 20% increase in energy efficiency compared to 1990 levels. Nuclear fission funding is allocated through the Euratom arm of Horizon 2020 which has already funded academic MSR projects across Europe. Funding rounds do not appear often and take over a year to establish. Sums of money are available for research and technology development projects but not development of prototypes.

KIC InnoEnergy is a European Commission program through the European Institution of Innovation & Technology (EIT) which has a primary goal of enhancing clean sustainable energy production. It is competition based with branches across Europe. Its entrance criteria are that the technology must have a Technology Readiness Level above five and a revenue stream available within five years. This rules out funding of advanced nuclear.

Research Councils have substantial annual funding at £3bn per year. This is not intended for industry and must go to research. The funds are distributed by seven councils. The most suitable for prototype development would be the Science and Technology Facilities Council (STFC) but this would be purely academic and current annual budgets are not sufficient to make substantial developments in advanced nuclear.

Direct Government funding, according to current policy is difficult as industry is intended to lead the way within the civilian nuclear power sector. Direct government funding would enable a faster route to development of advanced technologies. Even with extensive private investment, tangible government support is a necessity. The *Office of Nuclear Regulation*[▲] is fully occupied with decommissioning and the current new build light water programme; putting the assessment of a demonstration MSR higher on the ONR priority list would be a requirement.

Government support would invoke interest in industry. The authors have contacted several of the leaders in the nuclear industry and the response is common – if Government leads, industry will follow.

The cost to the public purse for full government commitment to this innovative agenda, if this process can be pursued until it is industry-standard, would be large. However, to put this in perspective, each

▲ *see Glossary*

one billion pounds spent, averaged over the UK working population (42,791,442 in 2015), is exactly £23.70. For a scenario of £50 billion over ten years this amounts to an economic cost of £117 per year per person. Not a trivial sum but arguably good value. This would provide a sturdy basis for the UK to become one of the international leaders in advanced nuclear development. There is little time for effective action but it has not passed.

SECTION OVERVIEW

- MSR technology has never been commercially available.
- Regulatory approval will be a lengthy, expensive process.
- Experimentation will be required for some new concepts and material applications.
- Obtaining funding is difficult due to the long commitment required and high risk of implementing a disruptive technology in a highly regulated environment.

7 Nuclear Regulation

The UK regime for nuclear regulation is intended to be flexible and proportionate and is based upon the principle of reducing risk to 'As Low As Reasonably Practicable'; it adopts a goal-setting approach rather than a prescriptive one. The legislation, guidance and safety and environmental requirements draw no explicit distinction between types of reactor (e.g. power production, test and research); rather, the onus is on the designer or operator to demonstrate how it achieves the necessary safety and environmental standards. While this approach results in a great deal of flexibility, it can make it difficult for the designer to determine the likely acceptability of novel designs.

The primary UK regulators of nuclear operations are:

Safety, Security & Transport

- the *Office of Nuclear Regulation*▲ (ONR) regulates the safety and security of nuclear facilities, transport of materials and ensures compliance with international safeguard obligations throughout England, Scotland and Wales; and

Environment

- the *Environment Agency*▲ in England
- the Scottish Environment Protect Agency (SEPA) in Scotland
- Natural Resources Wales in Wales

These environment agencies regulate discharges and disposal of radioactive waste from nuclear licensed sites and also the disposal of hazardous wastes. There are Memoranda of Understanding between the regulators to aid consistency of approach and prevent conflicting requirements being placed upon operators.

Legal requirements for a prototype MSR

As the law currently stands any nuclear reactor, whether prototype or commercial would be subject to:

- The Ionising Radiations Regulations 1999 – sets Statutory limits for dose uptake and invokes the principle of "So far as is reasonably practicable"
- Radiation Emergency (Preparedness and Public Information) Regs 2001 – requires an assessment of offsite consequences and may require an off-site emergency plan
- The Nuclear Installations Act 195 (as amended) – would require licensing by ONR and compliance with a suite of licence conditions
- The Environmental Permitting Regulations (in England and Wales) – Environmental Permits for aerial, liquid and solid disposal of radioactive and hazardous waste (e.g. beryllium)
- Planning considerations
- Justification – led by DECC and determined by Secretary of State
- Energy Act – Assessment and funding of decommissioning costs

There is no *de minimus* size or capacity for a nuclear reactor below which a nuclear site licence or environmental permit is not required.

▲ *see Glossary*

Discussion

Licensing and Authorisation of a new site would inevitably attract the need for public consultation by the relevant Environment Agency – and probably public body notification by ONR, in addition to any consultation required under the relevant planning regime.

New reactors have typically been licensed and permitted as part of the development of existing licensed sites. There is therefore little experience of licensing new sites and it is difficult to predict how proposals for a new licenced site for a demonstration plant might be received.

ONR and the environment agencies have licensed and permitted a wide range of reactors, but have not regulated any prototype/experimental reactors through design, construction into operation (all the UKAEA's reactors were shut down by the time of its licensing in 1991). The current strategy for regulators' engagement on new reactor technology is focused on those which a vendor or potential operator puts forward as complete designs, suitable for commercial deployment. Workload is forecast using a New Build Confidence Criteria[12].

The regulators have encouraged designers and potential operators to engage with them early in the design and development process to minimise regulatory risks and this has proved successful in the context of the Generic Design Assessment process. But this process for evaluating a new reactor proposal is designed for essentially complete designs, which have support for commercial deployment in the UK. The regulators are considering their strategy for research to support future regulatory work on novel designs but given the competing demands on their resources have not yet engaged widely in this area. There is currently little provision for institutional support for prototype or demonstration ventures in the UK with no explicit process for Small Modular Reactors or other new and emerging technologies.

It would be helpful for both regulators and designers to be able to engage early in the process to identify areas that may challenge the regulatory system and may necessitate skills not currently available to the regulators.

The skills needed for the regulators to support current technologies are predictable, but novel reactor designs may drive a need for new specialist skills (e.g. in the context of this report, *molten salt*[▲] chemistry). The availability of such skills may be very limited, making it difficult to obtain independent advice. Several universities such as Leeds could contribute to this shortage with their expanding involvement in nuclear engineering.

SECTION OVERVIEW

- No experience exists for licensing a prototype reactor or a new site.
- Regulatory burden for innovative technology is responsibility of the vendor.
- True innovation is severely restricted by the current process.

▲ *see Glossary*

8 Site Selection

As part of the government policy statement DECC carried out an assessment of eight potential sites[13]. Atkins preceded this with an assessment of alternative sites to the DECC report[14]. The siting selection of a prototype *molten salt*▲ reactor would have less restrictive technical constraints than those imposed in these siting assessments.

The prerequisites for implementation of a prototype reactor include:

- a suitable site on which to host it; and
- an organisation with suitably qualified and experienced staff to operate it.

The UK regulatory regime requires that any installation must be installed and operated on a site licensed under the Nuclear Installations Act 1965 (as amended) and disposals and discharges of waste should be in accordance with Authorisations/Permits issued by the relevant *Environment Agency*▲. It is many years since a new site was licensed and it is unlikely that a new site could be brought to availability in a realistic timescale. An existing licensed site would already have much of the necessary infrastructure in place and be operated by an organisation with experience and competence in nuclear operations.

Several of the existing sites are in commercial operation and it seems unlikely that the operators would welcome the distraction of a prototype reactor on the site, particularly one of such different technology. Equally it seems unlikely that any of the defence sites would be made available for development of prototype commercial reactors.

There have been no operational experimental/prototype reactors in the UK for some years, nor is there any UK organisation that is charged with investigating future reactor systems with bulk quantities of fissile material. The Nuclear Decommissioning Authority has responsibility for many of the UK nuclear licensed sites, formerly operated by UKAEA, but its focus is on their clean-up.

The NDA's current priorities are stated as:

- delivering and accelerating the work on the Legacy Ponds and Silos at Sellafield
- working towards the closure of the Dounreay site
- optimising routes towards Care and Maintenance across the Magnox fleet and at Harwell and Winfrith
- continuing Magnox fuel reprocessing in line with Magnox Operating Programme (MOP9)
- implementation of the Low Level Waste strategy at the Low Level Waste Repository (LLWR)
- supporting SME organisations by increasing overall spend with them to 20%
- removing all fuel from Sizewell A

The NDA has no remit in developing new reactor systems except in so far as it may assist site clean-up or storage or disposal of fuel and waste. But as clean-up progresses it is freeing land which cannot be released for unrestricted use for many years but might instead be made available for other nuclear uses. This would be a potential income stream for NDA for land of little monetary value.

The site operators implement the strategies defined by the NDA and have milestones to achieve which heavily influence their rewards. A parent body organisation or site licence company would therefore not embark on a major new scheme without the agreement and direction of the NDA.

▲ *see Glossary*

To be of interest to the NDA any proposal would need to:

- align with the NDA priorities outlined above; and
- not hinder the site decommissioning and clearance programmes.

The sites which could be suitable for siting a prototype MSR are within the remit of the Nuclear Decommissioning Authority: the Magnox sites as they reach the quiescent phase of the decommissioning plan; the chemical plant complexes (Sellafield, Dounreay, Harwell and Winfrith).

SECTION OVERVIEW

- The UK has no facilities for demonstrating new reactor technologies.
- The development process and timeline will be greatly simplified if an existing licensed site can be used.
- MSRs that burn plutonium could be of benefit to the NDA which owns some suitable sites.

Glossary of Terms

Accelerator Driven Systems (ADS)

Proposed accelerator-driven sub-critical reactor systems for nuclear energy generation employ a high energy proton beam that strikes a heavy element target to yield neutrons by spallation. The spallation neutrons then drive self-terminating fission chains in a sub-critical core which could be a molten salt. For a typical ADS proposal each energetic proton creates 20–30 neutrons from a thorium, mercury or other heavy metal target. The driver is planned as a 1 GeV, 10 mA current (i.e. 10MW) linear accelerator.

Described as one of the largest linear accelerators yet, the one at Oak Ridge National Laboratory is expected to reach as much as 3MW energy at target. It has a large footprint and operating team. Provision of an appropriate, reliable, and affordable linear accelerator is an unresolved challenge for ADS. Whether a useful ADS will ever happen is uncertain. It is not considered an option for the present study.

Denatured Fuel

Denatured uranium is when sufficient quantities of uranium isotopes that are non-fissile are mixed with fissile uranium to render it unsuitable for weapons proliferation. This is typically a high quantity of U-238 mixed with U-235.

Environment Agencies

Environmental aspects are regulated under the Environmental Permitting Act (or in Scotland the Radioactive Substances Act). The Agencies impose limits on the amount and types of material that can be disposed of to the environment (aerial, liquid and gaseous); conditions of the permits set out a framework within which the operator must operate. The limits and conditions are tailored to the plant and will be reviewed regularly throughout its lifetime.

The environment agencies set conditions that state that operators are required not only to comply with numerical limits on the levels of activity that may be discharged, but also to use 'Best Available Technologies' (BAT) (England and Wales) or 'Best Practicable Means' (BPM) in Scotland, to minimise the amount of radioactivity discharged. Operators are required to use BAT or BPM to minimise the volume and activity of:

- radioactive waste produced, which will require ultimate disposal under the environmental permit or authorisation;
- radioactive waste disposed of to the environment; and
- radioactive waste disposed of by transfer to other premises.

The permits may apply to the total amount of radioactivity that may be discharged by the site but will also place subsidiary limits on individual plants on the site.
The agencies consult publicly before issuing new or revised permits.

Fertile Material

A material, which is not itself fissile (i.e. fissionable by neutron bombardment), that can be converted into a fissile material by irradiation in a reactor. There are two basic fertile materials: uranium-238 and thorium-232. When these fertile materials capture neutrons, they are converted into fissile plutonium-239 and uranium-233, respectively.

Fission

The splitting of an atom releases a considerable amount of energy (usually in the form of heat) that can be used to produce electricity. Fission may be spontaneous, but is usually caused by the nucleus of an atom becoming unstable (or "heavy") after capturing or absorbing a neutron. During fission, the heavy nucleus splits into roughly equal parts, producing the nuclei of at least two lighter elements. In addition to energy, this reaction usually releases gamma radiation and two or more daughter neutrons. Mass lost as a result of fission is the source of the energy according to $E=mc^2$.

FLiBe

This is a eutectic mixture consisting of lithium fluoride (enriched in lithium-7) and beryllium fluoride. This is commonly employed in MSR designs.

FLiNaK

This is a eutectic mixture consisting of lithium, sodium and potassium fluorides.

Fluoride High-temperature Reactor (FHR)

The Fluoride High Temperature reactor is a solid fuelled, liquid salt cooled reactor, currently at a further stage of development than any of the liquid-fuelled configurations. It is being pursued by the Chinese as a step towards the liquid-fuelled version molten salt reactor and for building their knowledge and skill base as development progresses. The fuel is typically in a pellet or pebble form and has potential commercial use in high temperature applications.

Freeze valve

A section of pipe consisting of salt in solid state. As the temperature increases the salt melts and opens the 'valve'.

Hastelloy N

A nickel chromium alloy developed at Oak Ridge National Laboratory for the molten salt reactor programme. It has a high resistance to corrosion at high temperatures in an irradiated environment.

Molten Salts

Molten salts, usually liquids at 500°C and above, are a unique medium, consisting entirely of ions, known as cations and anions, positively and negatively charged species in equal proportions. They are a preferred candidate as a heat transfer medium as they have a high heat capacity, low viscosities and are chemically stable and non-flammable. Molten salts are stable at high temperatures and have low vapour pressures. The chemical compatibility between the salt and the structural material is determined by the chemical properties of the salt and mixtures thereof and redox potential. A wide range of salts exists which have been examined in detail at Oak Ridge National Laboratory. There is access to methods and data bases that are usually fit for purpose. Some of the parameters for modelling can also be estimated by methods developed by computational chemists. Phase diagrams have been produced for most common combinations which can be viewed here[15] along with recent accounts of the salt selection process for the Advanced High Temperature Reactor being developed between the United States and China[16].

Fluoride is the most common anion for molten salt selection. It has a minimal effect on neutron economy. Fluorine is the most electronegative element in the Periodic Table. From work at Oak Ridge National Laboratory most molten fluoride chemistry is well known and controllable. Three important chemical concepts play a major role in controlling the behaviour of liquid fuels, namely their solution chemistry, redox chemistry, and chemical activity. Solubility is of concern in keeping components in a homogeneous molten solution during reactor operation. Redox control is of significance in managing corrosion of containers and components.

Much effort has been made in the development of suitable alloys such as *Hastelloy N*[*]. The Oak Ridge National Laboratory reported a modification thereto that made it fully resistant to hot corrosion under radiation in an MSR. Chemical activity in the molten salt system was not of great concern during initial MSRE development. Today it is a topic of importance with the wider choice of fuels. Chloride salts are less commonly chosen because, amongst other reasons, sulphur is a decay product. Chemically, chlorides may be a better option in certain circumstances.

There is wider choice for cation selection. At Oak Ridge National Laboratory attention was, and still is, focused on beryllium and lithium. Lithium is available in two stable isotopes, Li^6 and Li^7. Li^6 must be avoided as it is a strong neutron absorber and produces unacceptable amounts of tritium. Li^6 is the most abundant form and the percentage of Li^7 must therefore be increased by enrichment, a simple process but at present no longer available on a commercial scale. With a chloride salt sodium can be beneficial as a cation. There is minimal tritium production but chlorine-36 and sulphur production must be managed.

Nuclear Fusion

Nuclear fusion occurs when elements of low atomic weight, such as hydrogen, helium, lithium, etc, fuse to form an atom with higher atomic weight. This can release huge amounts of energy ($E=mc^2$). It is the energy source for the sun and for the 'hydrogen' bomb. Harnessing this energy industrially would allow unlimited energy resources with little or no waste products.

The largest fusion reactor currently under construction is the €16bn ITER facility in Cadarache. It comprises a 60 m high 'Tokomak' tower holding the fuel, a deuterium-tritium hot plasma, retained in place by a very high magnetic field. It aims to produce 500 MW with 50 MW input power. The Tokamak, a spherical containment for plasma was developed in the USSR in the 1960s. Recent proposals are for a miniature version, a Lockheed Martin project, aiming to be complete by 2019 and capable of delivering 100 MW.

In 2010, Nobel Prize winner, Georges Charpak together with two colleagues, drew attention to three major obstacles to success for ITER, describing it as 'an ancient dream'. The obstacles are: an ability to maintain the unstable plasma in place; to produce tritium (12.3 years half-life) in industrial quantities; and to invent materials of construction for the high vacuum containment of thousands of cubic meters of hot plasma. The ITER project is aiming to solve the first of these problems by 2019.

Office of Nuclear Regulation (ONR)

ONR regulates safety through the Nuclear Installations Act 1965 (as amended) (NIA). This act requires an operator to hold a license to install, operate, decommission, etc. certain types of installation (e.g. nuclear reactors). Regulations made pursuant to the NIA prescribe activities which form part of the nuclear fuel cycle (e.g. enrichment, fuel fabrication and components of reprocessing).

The nuclear licensing regime does not extend to the Crown, but it does apply to sites operated on its behalf by a private organisation (e.g. Aldermaston). The UKAEA was exempted from the licensing provisions of the NIA until 1991 when its sites (Dounreay, Winfrith, Harwell and sites adjoining Sellafield and Springfields) were licensed, by which time all the reactors had ceased operation.

ONR's main tool in assessing the adequacy of installations is its Safety Assessment Principles: published in 1979 they have been updated to reflect changing standards and evolving techniques and have been benchmarked for consistency with IAEA Requirements and Standards.

The Safety Assessment Principles were originally devised for nuclear reactors but, as ONR's sphere of regulation has expanded, so these principles have been revised to ensure they are applicable to other relevant activities – e.g. fuel manufacture, enrichment and reprocessing. It is not necessary for any facility to meet the requirements of all the Safety Assessment Principles, but the safety case is expected to justify any exceptions.

ONR attaches conditions to nuclear site licences and over the years has developed a suite of 37 standard conditions; these are typically goal-setting, requiring the licensee to make and implement adequate arrangements for a range of key topics. But before granting a nuclear site licence ONR will need to satisfy itself that the prospective licensee has the necessary resources (financial and technical) to be able to discharge its duties under the licence.

Thermal Spectrum

The kinetic energy range of about 0.025 eV is the energy corresponding to the most probable neutron velocity at a temperature of 17 °C. After a number of collisions with nuclei in a medium (neutron moderator) at this temperature, neutrons are now at about this energy level, provided that they are not absorbed. Neutrons have a different and often much larger effective neutron absorption cross-section for a given nuclide than fast neutrons, and can therefore often be absorbed more easily by an atomic nucleus, creating a heavier, and often unstable isotope of the chemical element as a result (neutron activation).

Thorium

Primarily, three long-lived naturally occurring radioactive elements occur in the earth's crust, namely potassium-40, thorium-232, and two uranium isotopes, uranium-235 and uranium-238. Thorium, with a half-life of three times the age of the planet, is 3 to 4 times more abundant in the earth's crust than uranium. Uranium-235 is the only fissile material found in nature, and comprises 0.7% of naturally occurring uranium.

Thorium-232 and uranium-238 are fertile, that is they convert to fissile material by breeding, achieved by absorbing a neutron. The thorium transmutes to uranium-233 and the uranium to plutonium-239, both fissile fuel sources. Thorium is relatively inconvenient to use as solid fuel, but has many advantages in thermal spectrum liquid-fuelled reactors. These include:

- the decay chain for uranium-233 (bred from thorium) comprises short-lived actinides with advantages in terms of waste disposal;

- liquid-fuelled reactors burn most of their fuel, leaving little waste, whereas solid fuelled reactors only burn a few percent of the fuel;

- thorium is not costly to acquire, partly because it occurs as a single isotope so needs little preparation for use;

- thorium used in liquid-fuelled reactors can provide planetary energy requirements for tens of thousands of years to come.

An application of thorium in our environment is its contribution to the production of heat within the earth's core. The figure below shows that thorium's radioactive decay heat produces more heat than other elements in the earth's core and is the most stable source of that heat.

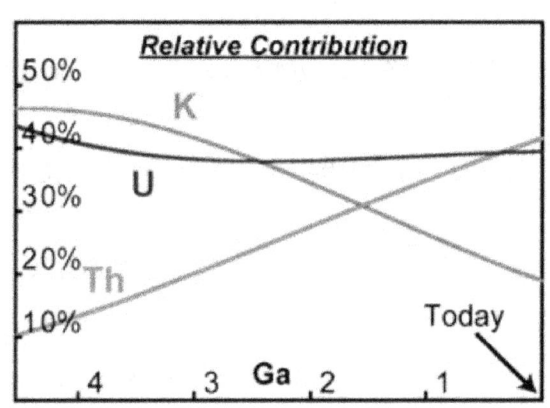

Decay heat produced by various elements within the earth's core over time. Ga = giga-annum. It is estimated that radioactive heat produces around half of the heat beneath the earth's crust[17].

References

1   Department for Business, Innovation and Skills (2013), *Nuclear Industrial Strategy -The UK's Nuclear Future* http://tinyurl.com/cwa9xqu

2   Department for Business, Innovation and Skills (2013), *Ad Hoc Nuclear Research and Development Advisory Board* http://tinyurl.com/qbuewqm

3   Department for Business, Innovation and Skills (2013), *Nuclear Industrial Vision Statement* http://tinyurl.com/chuw47h

4   Department for Business, Innovation and Skills (2013), *Long-term Nuclear Energy Strategy* http://tinyurl.com/bn49olr

5   Department for Business, Innovation and Skills (2013), *A Review of the Civil Nuclear R&D Landscape in the UK* http://tinyurl.com/mfnu3qy

6   Department for Business, Innovation and Skills (2013), *Nuclear Energy Research and Development Roadmap: Future Pathways* http://tinyurl.com/mjnxejl

7   Presentation on the Generic Design Assessment by Colin Potter, ONR and Alan McGoff, EA. December 2013.

8   A.M. Weinberg, *The proto-history of the molten salt system*, A speech to the Korean Scientific Delegation at ORNL, 23rd February 1997 http://tinyurl.com/olegm3t

9   Oak Ridge National Laboratory-3936, *MSR Experiment Semiannual Progress Report*, February 1966.

10   OECD/NEA-5427 - *Pyrochemical Separations in Nuclear Applications, A Status Report*, 2004.

11   W. De Kruijf, *Reactor Physics Analysis of the Pin-Cell Doppler Effect in Thermal Neutron Reactors*, PhD at Technical University Delft, 1994.

12   Office Nuclear Regulation, *Board Meeting Minutes*, 3 February 2015 http://www.onr.org.uk/meetings/2015/onr-15-02-05.pdf

13   DECC report, *National Policy Statement for Nuclear Power Generation (EN-6) Volumes I & II* (2009).

14   Atkins report, *A consideration of alternative sites to those nominated as part of the Government's Strategic Siting Assessment process for new nuclear power stations* (2009).

15   Oak Ridge National Laboratory-2548, *Phase Diagrams of Nuclear Reactor Materials*, 1959.

16   Oak Ridge National Laboratory/TM-2006/12 *Assessment of Candidate Molten Salt Coolants for the Advanced High-Temperature Reactor (AHTR)*, 2006.

17   R. Arevalo Jr., W. McDonough, Ma. Luong, *The K/U ratio of the silicate Earth: Insights into mantle composition, structure and thermal evolution*, Earth and Planetary Science Letters vol.278 pgs 361–369 (2009)

Appendix A

MSR Activity Today

Contents

1 Discussion

MSR activity is to some extent defined by civil nuclear spending on research and development. As discussed previously, the UK investment in civil nuclear R&D is in the tens of millions of pounds compared to the billions in Japan. The larger part of this UK funding is found to be for nuclear fusion. For liquid-fuelled MSR technology – the topic of the present study - R&D is a very small proportion of the total annual R&D spend. For liquid-fuelled MSR technology there is no apparent state support for implementation. The exception to this pattern is China.

Internationally the MSR activity scene, with the exception of China, is fragmented and these fragments are small and some may be easily overlooked. Six liquid-fuelled MSR proposals have been selected for consideration for this study. There is now increasing public interest in MSRs – some of Kirk Sorensen's YouTube MSR presentations have had over two million hits. If led by government funding directed to MSR implementation, this sector is ready to grow rapidly – again as evidenced by activity in China detailed further on.

The Generation IV Forum, of which the UK was a founder member but has since been relegated to observer status because of lack of any UK advanced reactor research - was set up in 2000. It selected six advanced reactor systems and coordinates their international activities. The MSR was selected as one of the six. While there are a number of small projects around the world on the use of molten salts in fuel processing, actual molten salt reactor design programmes are rare. The Molten Salt Fast Reactor (MSFR) developed at Grenoble has particular significance. It was recently selected by the Generation IV International Forum (GIF) as the liquid-fuelled reactor reference model.

The MSFR project embraces an extremely advanced molten salt reactor, a fast spectrum single fluid thorium breeder with on line reprocessing. This is technically very challenging and is therefore essentially an academic exercise only. Because of its technical challenges it is seen as the most difficult project with the longest time to commercialisation of all the Gen IV reactors. The other significant international project, among the total of six selected as options, is a collaborative effort between China and the Oak Ridge National Laboratory which only addresses the simpler challenge of a molten salt cooled but conventionally fuelled reactor. This is known as the *Fluoride High-temperature Reactor (FHR)*[▲].

MSR Resurrection

The analysis of the Gen IV Forum, in considering molten salt reactor concepts correctly evaluated the scenario. However, in the 10 years since that analysis was undertaken, the technology landscape relating to molten salt reactors has utterly changed and these changes have originated in small, entrepreneurial start-up companies, rather than in large government laboratories.

Three principal changes in design philosophy have taken place in those 13 years since the original Gen IV selection process in 2002. The first of these is to design simpler, less ambitious, molten salt reactors that do not breed new fuel, do not require online fuel reprocessing and which use the well-established enriched uranium fuel cycle. The second is to design small modular molten salt reactors with lower capital costs and short design lifetimes. The third is a more radical design simplification that eliminates all pumped circulation of the molten salt fuel.

These new designs can be moved relatively quickly to the detailed design, approval and prototyping stage. New science is not required, merely disciplined development effort. Molten salt reactors have thus moved from the back of the technology pack to the very front. No major company or national government has yet grasped this opportunity, apart perhaps from China. In contrast to all the other Gen IV reactor concepts, the opportunity for global leadership and intellectual property ownership still exists. This window of opportunity will not remain open for long.

▲ *see Glossary*

A2 MSR Designs Reviewed

MSR Proposal Technical Reference Table

Organisation	Flibe	Martingale	Moltex
Reactor Name	LFTR	ThorCON	SSR
Spectrum	thermal	thermal	fast
MWth Pilot	2	8-550	150
MWth Commercial	2225	550 (modular)	2500
Fuel			
Fuel Salt	Uranium & Thorium in LiF - BeF2 - (Th/U)F4	Uranium & Thorium in NaF - BeF2 - (Th/U)F4	SNF in NaCl - (U/Pu/La)Cl
Coolant Salt	NaBF4 - NaF	NaF - BeF2	ZrF - KF - NaF
Reprocessing			
Gaseous FP removal	assisted	assisted	off gases in space above
Other FP removal	bismuth FP extraction	off site FP removal	plate out on vessel walls off site FP removal
Other processing	Pa/U-233 separation (from Th)	off site uranium separation	-
Novel Safety Features			
	freeze valve /drain tank	freeze valve /drain tank	sealed unit
	shutdown rod	sealed primary loop	water cooling surround
		continuous passive decay heat removal	shutdown rod
		shutdown rod	poison pill
Materials Proposed			
Moderator	graphite	graphite	N/A
Vessel	Hastelloy N	SS316Ti	SS316
Primary Piping	Hastelloy N	SS316Ti	PE16, (or Molybdenum)
Secondary Piping	Hastelloy N	SS316Ti	SS316

MA - minor actinides

La - Lanthanides

FP - Fission Product

SNF - Spent Nuclear Fuel

MSR Proposal Technical Reference Table (continued)

Organisation	Seaborg	Terrestrial	Transatomic
Reactor Name	SWaB	IMSR	TAP
Spectrum	thermal-epithermal	thermal	thermal-epithermal
MWth Pilot	50	80	20
MWth Commercial	250 (modular)	80 to 600	1250
Fuel			
Fuel Salt	Thorium & SNF in LiF - (Th/Pu/MA)F4	Uranium in NaF - RbF - UF4 (or LiF-BeF2-UF4)	Uranium in LiF - (U/Pu/La)F4 SNF in later editions
Coolant Salt	LiF - NaF - KF	KF-ZrF	LiF - KF - NaF
Reprocessing			
Gaseous FP removal	assisted	off gases in space above	assisted
Other FP removal	bismuth FP extraction	off site FP removal	off site bismuth FP extraction plate out on nickel filter
Other processing	decladding of SNF pellet uranium separation from SNF thorium level control	-	-
Novel Safety Features			
	overflow tank freeze valve /drain tank fuel coolant (twice around core) shutdown rod	buffer salt zone sealed unit water cooling surround buoyancy shutdown rod poison pill	freeze valve /drain tank passive air cooling shutdown rod
Materials Proposed			
Moderator	graphite	graphite	zirconium hydride (clad)
Vessel	Hastelloy N / SiC	Hastelloy N - tbc	SS316 + Hastelloy N
Primary Piping	Hastelloy N / SiC	Hastelloy N - tbc	SS316 / Hastelloy N
Secondary Piping	Ni-201	unknown	SS316 / Hastelloy N

MA - minor actinides

La - Lanthanides

FP - Fission Product

SNF - Spent Nuclear Fuel

MSR 1 – Flibe Energy – Liquid Fluoride Thorium Reactor (LFTR)

Flibe Energy, one of the first to resurrect the molten salt reactor concept, and based in the USA propose a 2MWth two fluid breeder design. It is based on work carried out by the Oak Ridge National Laboratory team in the 1970s. It operates in the thermal spectrum moderated by graphite. Its fissile element is uranium-233 which is bred from thorium in a blanket salt at the outer edge of the reactor core. The thorium is converted to protactinium by neutron bombardment which is then separated by means of an online reprocessing system and added back to the inner core once the protactinium has decayed to uranium-233.

Both the fissile and fertile fuel solutions are composed of lithium-7 beryllium fluoride salts. These are pumped through the reactor core in the centre of hexagonal graphite logs. Figure 3 below shows a section through the core where the graphite logs containing fissile fuel can be seen in blue and those with fertile fuel in green. The black logs are reflectors to concentrate the neutrons within the core. The blanket fuel is taken to one reprocessing stream to separate the U-233 and remove fission products before being brought back into the core. The fissile fuel is separated for fission product removal and addition of the pure U-233 from the fertile stream. A graphite control rod exists at the centre of the core for minor alterations in reactivity. Major alterations to reactivity levels are carried out by adjustment of the composition of the fuel salt.

The blanket fuel salt flowing around the perimeter of the core helps to maintain the vessel walls at a constant temperature. Several drain tanks exist for the various salts where the fluid would flow in the event of any incident. It is envisaged that the fuel salt drain tank would require cooling by steam tubes through the salt. This is due to the decay heat which is common to all reactors. The tanks would be heated to keep the salt in a liquid state.

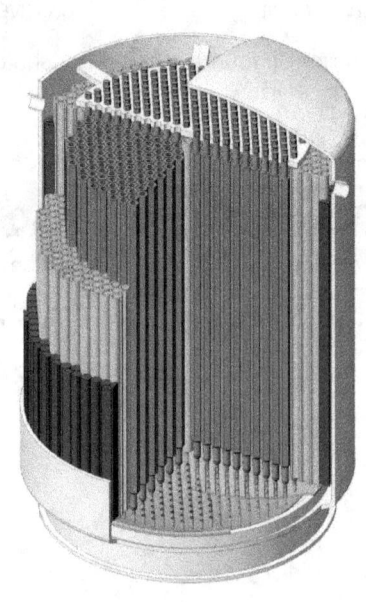

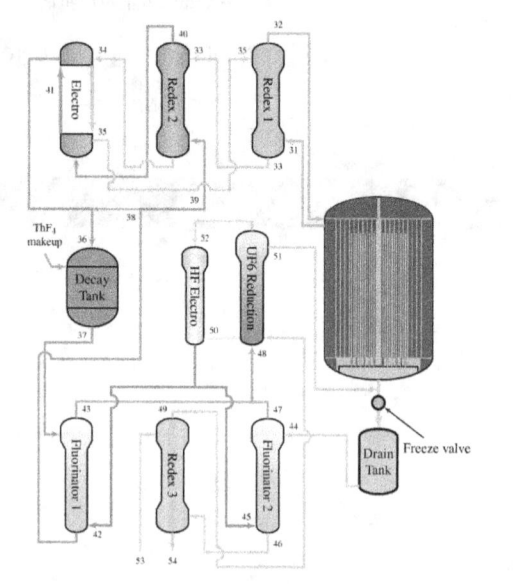

Figure A1
Illustrative view of the core showing the fissile zone in blue, fertile zone in green and the reflectors in black at the perimeter.

Figure A2
Schematic of LFTR reprocessing system

The reprocessing system is based on a reductive extraction system with a metallic bismuth and an electrolytic cell to separate the uranium and protactinium from the blanket salt. Once the protactinium has decayed to uranium only, it goes through a fluorination process and hydrogen reduction process before joining the fissile fuel salt at the bottom of the reactor core. The depleted fuel salt goes through a similar fluorination process for the fresh U-233 to be added and a reductive extraction process with bismuth metal for fission product removal. As with several other proposals, it is planned for the reactor and drain tanks to be located underground. The main material for salt containment in the pipework and vessels is *Hastelloy N*▲.

Being a breeder reactor the LFTR design utilises the full potential of thorium. It is a valid proposal which can be taken forward for development. Little experience exists with the specific reprocessing methods required which could potentially delay the development of a first-of-a-kind in the UK.

More information: www.flibe-energy.com

MSR 2 – Martingale Inc. – ThorCon

The ThorCon design is a single fluid thorium converter reactor that operates in the thermal spectrum. It is in principle, similar to the MSR Experiment and its fuel is *denatured*▲ using a combination of U-233 from thorium and U-235 enriched from mined uranium. Its core is graphite moderated and the full scale version is proposed to run at 550MWth. Several of these cores would make up a larger plant with each being replaceable after 4 years to minimise risk of critical component failure. A centralised facility is proposed to reprocess the spent fuel salt from multiple plants. The design team come from a shipping background and brought in nuclear expertise from members of the MSR community from across the United States. As a concept, a pilot-scale version of this plant would be similar to the MiniFuji, a concept that the Japanese have been working on for a long time, discussed later. Both designs are replicas of the MSR Experiment with minor alterations and no new technology. The review carried out for this study assessed both concepts primarily utilising the more available ThorCon data. ThorCon envisages a full scale prototype plant that steps the power level up in stages. The cost savings by building a smaller plant are not valued as high as the shortened programme going straight to full scale.

The fuel is a molten sodium beryllium fluoride with dissolved thorium and uranium tetrafluorides, specifically avoiding lithium-7 for schedule and cost reasons. The fuel salt is pumped through an array of hexagonal graphite logs surrounded by a boron carbide shell which protects an outer stainless steel vessel from excessive neutron bombardment. The salt then enters the heat exchanger at 704°C and leaves at 564°C. The primary coolant salt is a similar sodium beryllium fluoride which then transfers its heat to a secondary coolant salt, similar to that used in solar panels. The solar salt both traps any tritium and prevents any leak from the steam generator from getting to the fluoride salt.

Decay heat cooling of the reactor vessel is done by an array of vertical pipes of water that surround the vessel. The water circulates by natural convection to a storage tank at ground level. When more heat is removed the rate of convection is higher. A drain tank at the bottom of the core vessel is kept isolated by a fuse valve which opens in the event of overheating. There is no online processing of the salt. This is dealt with at the end of the reactor core's life. Off-gassing is dealt with by storage in several hold tanks at varying temperatures and pressures. It uses a helium technique to separate the gases from the salt as per the MSR Experiment.

▲ *see Glossary*

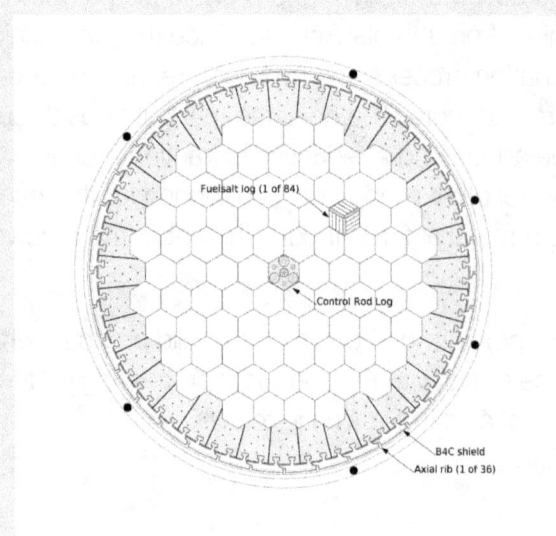

Figure A3
Section through core in plan view showing the graphite logs

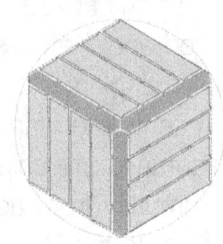

Figure A4
Close up section through graphite log

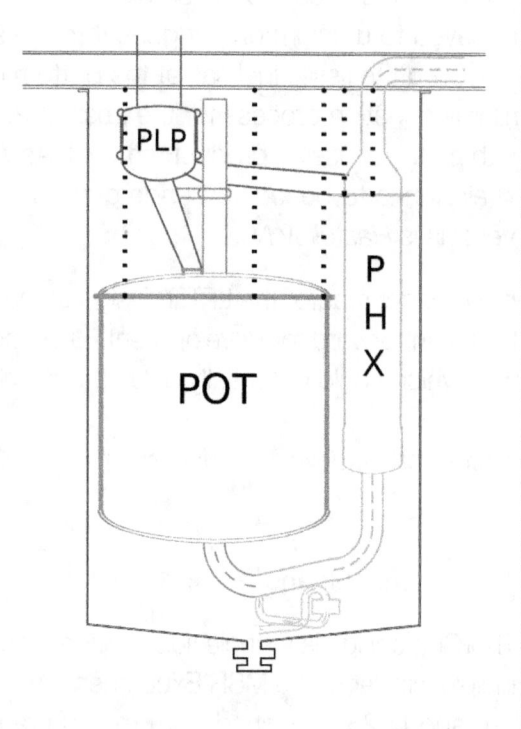

Figure A5
Section through reactor vessel showing the core - 'POT', pump - 'PLP and heat exchanger - 'PHX'

One major benefit of this design is that the basic concept is proven. The possibility to go straight to a larger plant is claimed if the experience from the MSR Experiment is used. The designer's proposal is to prefabricate 500 tonne components that get barged to the power plant site. This has the potential to dramatically reduce costs and improves the quality.

More information: www.thorconpower.com

MSR 3 – Moltex Energy – Stable Salt Reactor (SSR)

The Stable Salt Reactor has a design team based in the UK. It is a fast spectrum pool type reactor, again originally conceptualised at ORNL. The full size version is proposed at 1GWe and the prototype at 150MWth, large enough to reach criticality. It would operate at a much lower power. It is fundamentally different to the other proposals in that the fuel salt is static.

The fuel salt, comprised of 60% sodium chloride and 40% plutonium, uranium and lanthanide trichlorides sits in nickel-chromium alloy tubes (PE16) at the centre of a pool of coolant salt. These are arranged in arrays similar to conventional solid fuel arrays in a light water reactor. The primary fissile fuel is Pu-239 recovered from conventional reactor spent fuel. The coolant salt is a fluoride (ZrF_4-KF-NaF) which transfers the heat from the centre of the pool to the heat exchangers at the outside edge by natural convection, assisted by a small impeller. The heat exchangers are sufficiently far from the fuel tubes that neutron bombardment is negligible. The neutrons are stopped by a reflector at the outer edge of the fuel tube array and by the coolant salt itself. The fuel tubes have a lifetime of 5 years before they are moved to the outer edge of the pool for cooling.

The fuel has a strong negative coefficient of reactivity and requires no external cooling. The natural convection of the coolant salt will remove any decay heat in the event of a failure. This is assisted by water pipes around the perimeter of the tank. Shutdown rods exist to maintain the reactor subcritical when needed. In the event of failure of all heat removal/shutdown processes the coolant salt will boil, condense on the roof of the structure and flow back to the tank. At this high temperature, the fuel salt will still remain in a liquid state without boiling.

The production of the fuel would be by a multistage reductive extraction process using bismuth metal as it is immune to low levels of contamination. The coolant fluoride salt is more readily available and relatively cheap. Fission gasses are released at the top of the fuel tubes and into an argon space above the coolant salt. Xenon and krypton are left to decay to stable gases and other gases are compressed for storage until safe. The fuel tubes are designed such that other more problematic gases are maintained within the fuel salt or condense on the upper section of the tube. At the end of their life, the used fuel tubes will have no long lived actinides but will require storage for several years before being less radioactive than mined uranium.

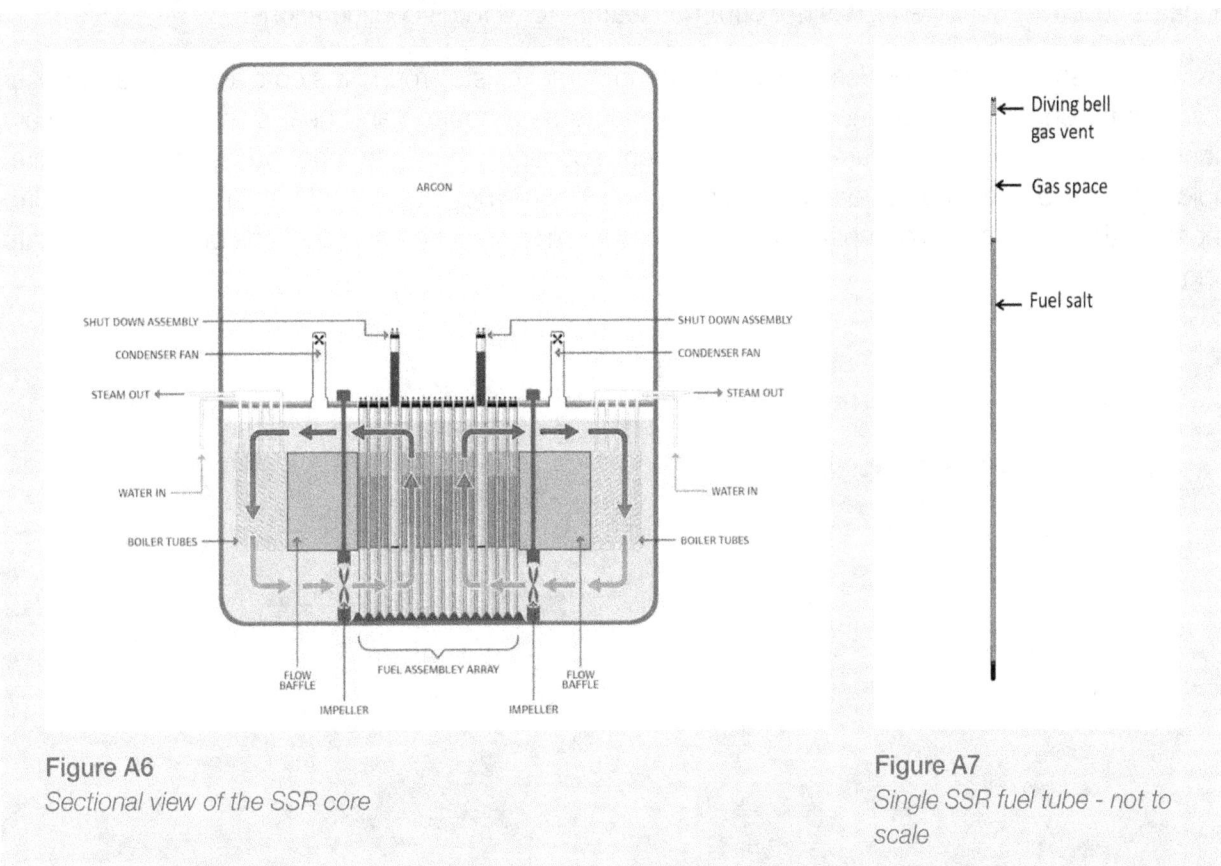

Figure A6
Sectional view of the SSR core

Figure A7
Single SSR fuel tube - not to scale

Being a static fuel, there is no need for pumps or other devices to control the flow. Fast reactors are more suited to burning the long lived troublesome actinides in spent nuclear fuel. Complications with the use of graphite are also avoided. As a concept, this proposal is deemed to be valid. Experimental data of the heat transfer properties of the fuel salt tube to coolant will be required to validate thermodynamic simulations.

More information: www.moltexenergy.com

MSR 4 – Seaborg Technologies – Seaborg Waste Burner (SWaB)

The SWaB prototype proposal is a 50MWth single fluid reactor that operates in the thermal-epithermal spectrum. It is graphite moderated and fuelled by a combination of spent nuclear fuel and thorium. The design team based in Denmark, is a combination of physicists and chemists from the Niels Bohr Institute and the Technical University of Denmark.

The fuel salt is a lithium-7 fluoride with dissolved thorium and fissile actinides from spent nuclear fuel. This is pumped through the graphite core and heat exchanger into a hold tank which is an important safety device. In most accident scenarios the fuel flows from this hold tank into a drain tank where it is rendered subcritical. A *'freeze valve'*[▲] also exists as a secondary safety feature. The core itself consists of hexagonal graphite columns through which the fuel salt flows in the centre. An operating temperature of 700°C is maintained by a secondary flow of coolant salt.

The reactor relies on a novel, patent pending, on-board chemical fluorination flame reactor, based on the Fluorex process which can continually extract fission products from the salt during operation. This reprocessing system can process spent fuel pellets from conventional reactors whilst removing the uranium for other uses. The system is also used to adjust the fuel levels in the salt, avoiding the need for control rods. Shutdown rods do exist in the core but primary reactivity control is via the coolant salt pump which changes the temperature of the fuel salt in the core thus altering reactivity due to its strong negative reactivity coefficient. This is a common feature to all MSR configurations.

The inner vessel is formed of Hastelloy N. A second barrier exists around the core and heat exchanger made of boron carbide clad in Hastelloy N. This acts as a neutron and heat shield and as primary containment in the event of an extreme accident scenario where the inner vessel fails. The core, hold tank and off-gassing system are contained within a Ni-alloy vessel (3rd barrier). Finally, the core, reprocessing system, and dump tank are contained within a steel dome under large blocks of concrete underground.

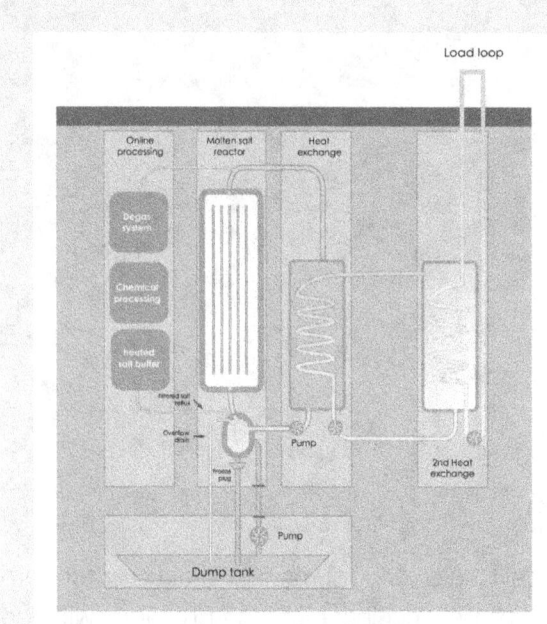

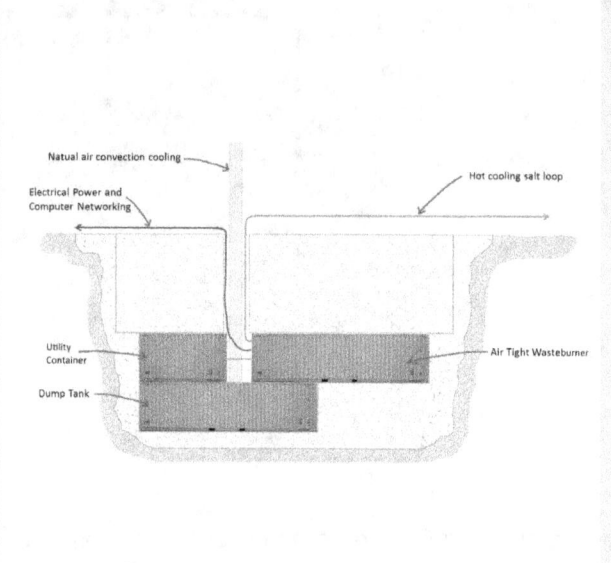

Figure A8
Schematic of reactor

Figure A9
Illustration of SWaB reactor in sealed container style unit under several concrete slabs

▲ *see Glossary*

This is a credible design with innovative features suitable for modularised construction. It is in its early stages of development with a competent and motivated design team. The reprocessing system brings many benefits but would delay the deployment of a first-of-a-kind pilot MSR. This is a suitable device for efficient utilisation of the spent nuclear fuel stockpiles. The concept is appropriate for further development.

More information: www.seaborg.co

MSR 5 – Terrestrial Energy – Integral MSR (IMSR)

The Integral MSR is also based on the MSR Experiment but has been modified to have a more sealed, passive approach. The core design team is based in Canada with international involvement and support. An 80MWth prototype reactor operating in the thermal spectrum is proposed.

The likely fuel will be sodium rubidium fluoride with potential to change to FLiBe, both bringing different advantages. With a graphite moderator inside a sealed unit it is designed to fit on the back of a lorry. This unit contains the fuel salt, moderator, heat exchangers and pumps. The pumps located at the top, circulate the fuel salt down through the heat exchangers and core. There are multiple pumps and heat exchangers to allow for breakdowns without affecting performance. A secondary coolant salt loop takes the heat away. The plant is fuelled with 5% low enriched uranium where the U-235 is *denatured*[▲] with U-238. This core is modular, designed at a high power density for replacement after a seven year cycle in a plant with a lifetime of over thirty years.

The core is moderately cooled by the fuel salt flowing around the perimeter as it circulates. A buffer salt zone exists around the entire core which is in solid state under normal operating conditions, and will gradually liquefy as decay heat is emitted. As it melts further, it displaces more heat by convection. Water filled cooling pipes around this buffer salt zone reduce the temperature further and maintain reduced temperatures for several weeks until which point temperature would be sufficiently low that normal radiative losses would suffice. Primary shutdown is via a buoyancy driven control rod kept outside the central core by positive flow and density difference. In the event of no circulation and/or increased temperature (reduced density) the rod will drop rendering the reactor subcritical. Secondary shutdown activation is provided by a neutron poison, released in the event the reactor exceeds a given temperature.

There is no online reprocessing. It is currently envisaged for gases to build up in the reactor over the seven year lifetime and left to decay along with the salt for a seven year cooling period at the end of the core's life. In the event that gases are excessive they can be taken off and bottled for storage. Fuel salt can be removed off site for reprocessing at the end of the cooling period or left in the reactor for long term storage.

▲ *see Glossary*

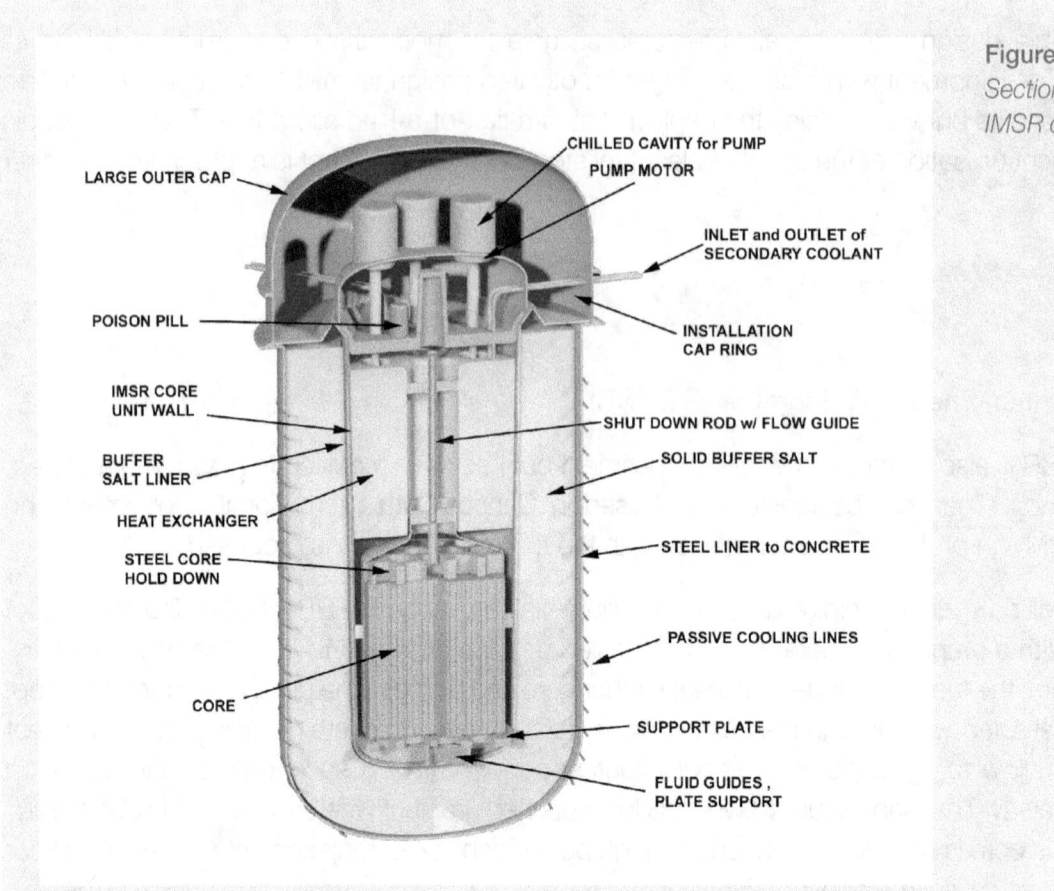

Figure A10
*Section through
IMSR core*

The Integral Molten Salt Reactor has advantages with its 'seal and swap' approach reducing on site complications and risks. Using low enriched uranium, it has a fuel cycle that regulators are familiar with. This proposal is deemed valid for developing fully and launching commercially. The design team are currently in discussions with Canadian regulators to progress their safety case.

More information: www.terrestrialenergy.com

MSR 6 – Transatomic Power Reactor (TAP)

Transatomic Power's proposed design is a 20 MWth demonstration reactor which is similar to the MSR Experiment except for its utilisation of zirconium hydride (instead of graphite) as a moderator and LiF-based salt (instead of a FLiBe-based salt). The design assessed was based on a 520MWe version. As a single fluid configuration, it operates in the thermal spectrum with a high neutron flux in the fast spectrum. This minimises neutrons in the epithermal region to minimize parasitic neutron losses. The design is from a start-up company based in the US that originated out of MIT from which it has maintained strong links.

The fuel is a 99.5% lithium-7 fluoride salt with dissolved plutonium, uranium and lanthanide trifluorides from conventional PWR spent nuclear fuel. This reactor is configured as a waste burner, annihilating far more troublesome long lived actinides than it produces over its lifetime. This is assisted by part operation in the fast spectrum. The initial versions would however be run on low enriched uranium. The vessel is constructed of 316 stainless steel with modified *Hastelloy N*[▲] and surrounded by a concrete structure. The novel moderator is less prone to expansion/contraction compared with graphite and is a

▲ *see Glossary*

more effective moderator allowing for a greater fuel-volume ratio at 50%. This permits a lower uranium enrichment down to 1.8%. It does however require cladding to separate it from the fuel salt. A coolant salt (LiF-KF-NaF) is used to transfer the heat to a steam generator.

A *freeze valve*[▲] exists at the bottom of the reactor vessel for plant shutdown or in an accident scenario. Cooling of the drain tank is provided by a passive ventilation stack that can dissipate the decay heat sufficiently quickly without any power input. Shutdown rods within the moderator can maintain the reactor in a subcritical state. A control rod exists for moderate regulation of reactivity but primary control is by altering the amount of heat removed as per other designs.

Fission products are removed batch wise and fresh fuel is added so that there is a constant fuel mass. Gasses are continuously removed, compressed and stored. Noble metal solid fission products are removed as they plate out onto a nickel mesh filter located on a side stream of the salt. Methods for removal of the dissolved lanthanides are under review but will likely involve a liquid metal/molten salt extraction process which turns the waste into oxide form.

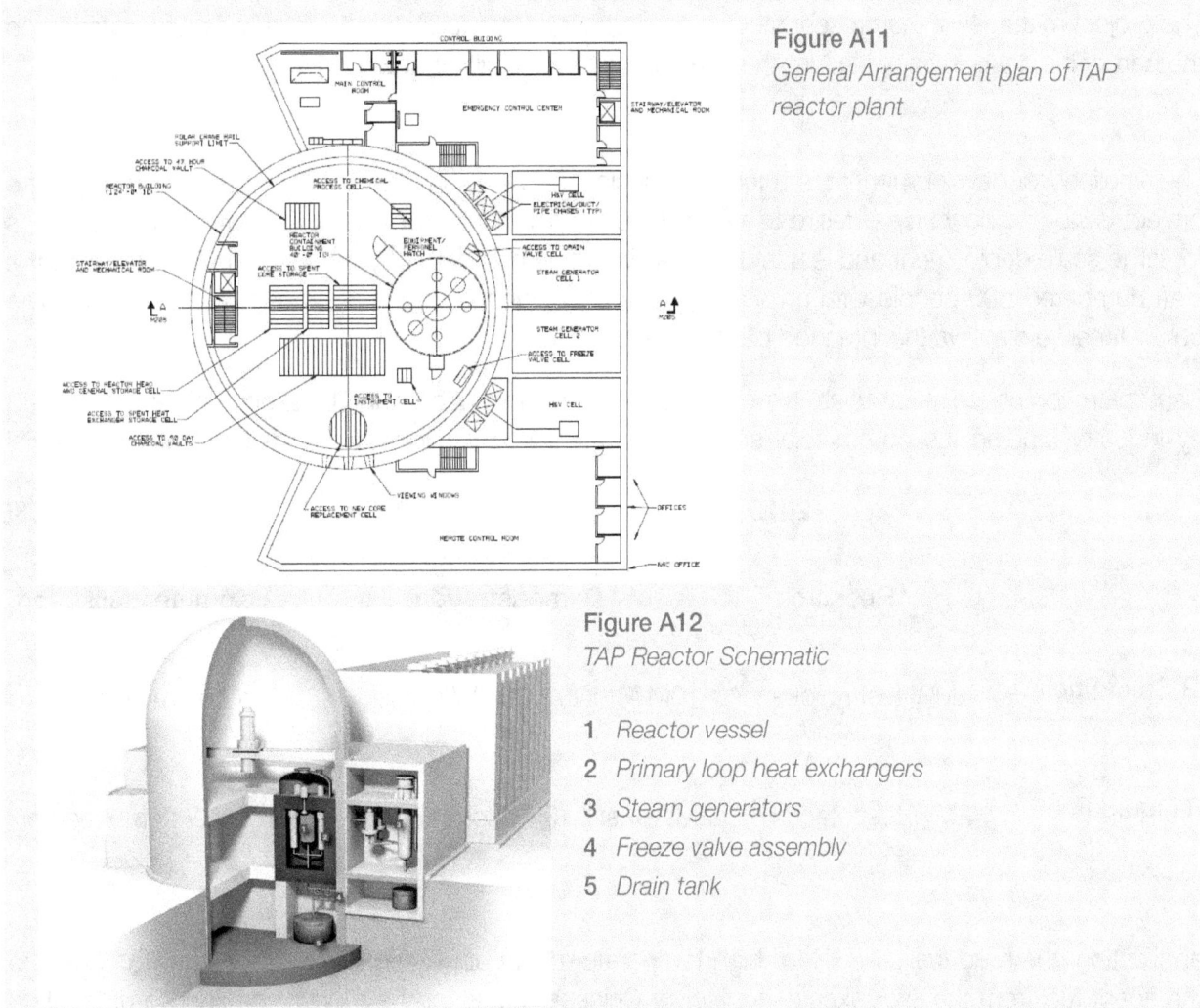

Figure A11
General Arrangement plan of TAP reactor plant

Figure A12
TAP Reactor Schematic

1 *Reactor vessel*
2 *Primary loop heat exchangers*
3 *Steam generators*
4 *Freeze valve assembly*
5 *Drain tank*

Experimental work is progressing at the moment on the use of zirconium hydride as a moderator and its cladding materials. The application of novel materials brings benefits, particularly its efficient utilisation of spent nuclear fuel but it also sets it at a longer time to deployment than more traditional configurations. This design is deemed valid and suitable for full development.

More information: www.transatomicpower.com

▲ *see Glossary*

A3 The Chinese Thorium Molten Salt Reactor project (TMSR)

Although of great interest the Chinese proposal would not be a feasible project for a UK pilot-scale demonstrator reactor, which is the central topic for this report. In January 2011 the Chinese Academy of Sciences set the development of a liquid-fuelled MSR as the top priority. The announcement stated 2015 as a target for a 2MW pilot MSR. The project was assigned to the Shanghai Institute of Applied Physics with a US$350 million budget. After some activities the current programme is for a change in top priority to a 10 MW solid pebble-fuelled salt-cooled demonstration reactor (*FHR*▲) perhaps in 2017. The 2MW liquid-fuelled MSR, similar to the ORNL MSR Experiment reactor, follows some years later. The project manager of the present study visited Shanghai in March 2015. There was no hesitation on the part of his hosts in showing him their capabilities and facilities at Shanghai.

TMSR Program

The timelines for development at Shanghai are unclear and they change as the project progresses. The project is slowed down by a desire to expand their experimental database. This lays the foundations for large scale deployment and is useful for training large numbers of young staff. In March 2015, the staff numbered 600 people with an average age of 31 who each have a steep learning curve. The below timeline displays the intended pace of development.

Demonstration phase reactors are expected to be funded by national and local governments, followed by industry funding the commercial scale versions. Three versions are envisaged for commercial

	2015	2025	2035
	Research	**Demonstration**	**Commercialisation**
MSR Solid Fuel	2MW test reactor	10MW test / 100MW demo	1GW commercial reactor
MSR Liquid Fuel	2MW test reactor	2MW test / 10MW online repro.	100MW demo reactor

application, the solid fuelled FHR for high temperature heat application, a waste burner to consume the spent fuel from the ever expending fleet of Chinese PWRs, and a thorium breeder to utilize their vast resources of thorium.

The budget is for the Shanghai Institute of Applied Science (SINAP) who run the project. It does not include any infrastructure costs such as the newly constructed research buildings or utilities. Other organisations are involved such as the Shanghai Institute of Organic Chemistry who are establishing methods for lithium production and the Shanghai Nuclear Engineering Research Design Institute who are developing the safety case for licensing.

▲ *see Glossary*

Long Term Commitment

Although there have been changes in the political situation lately the project will most likely continue for the following reasons:

- China does not have sufficient uranium resources for sustainability but does have thorium.

- The current energy capacity is 1500GW. Capacity of 2800GW is expected for 2030.

- Pollution levels are intolerably high in Shanghai and North East China. China has the highest CO_2 production rate per GNP in the world.

- Finding PWR sites is a challenge where available land with sufficient water resources is scarce. MSRs with a smaller footprint and less need for large volumes of cooling water are better suited.

The project reactor site has been chosen north of Shanghai. Local planning issues are causing delay. The reactor utilises a graphite core with a *FLiBe*[▲] salt. It has a heat decay removal system which operates passively by natural convection. The first version will be fuelled with uranium enriched to below 20% U-235. Nickel based alloys are being used for containment, pumps, and piping. They are experimenting with Hastelloy N but have developed their own similar version known as GH3535.

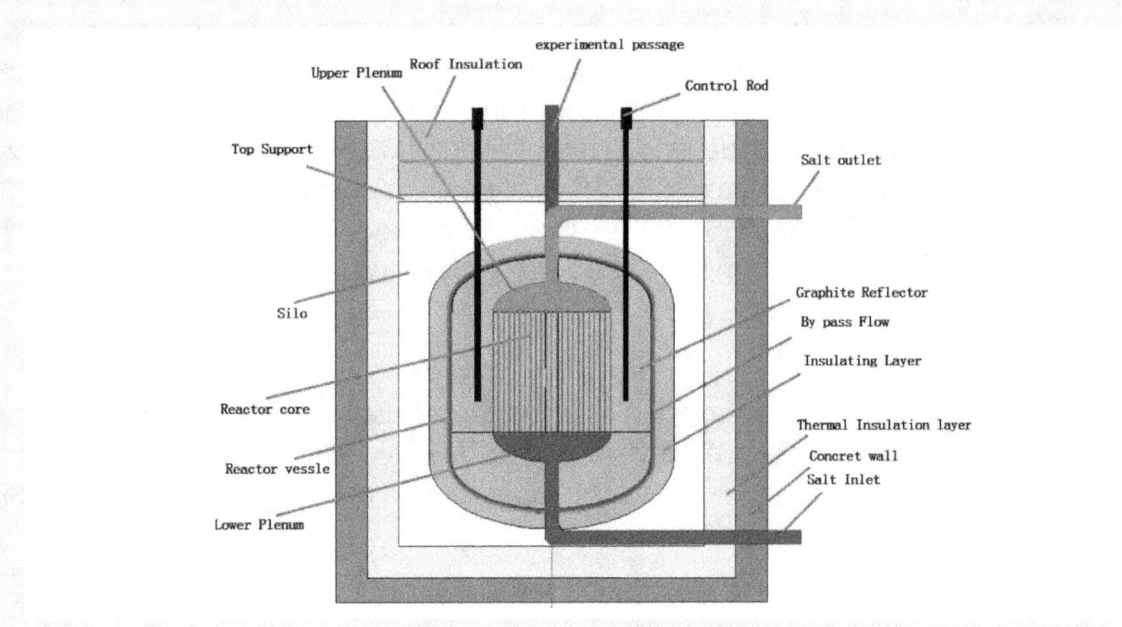

Figure A13
Main components of the Chinese Molten Salt Reactor Design

Current experiments at Shanghai include:

- water cooling loop to demonstrate passive operation of heat decay,

- enrichment of lithium-7 techniques,

- various techniques on salt reprocessing,

- various salt loops for pipe, pump and salt testing,

- thorium fuel and salt preparation,

- material performance and weld techniques for the nickel alloys,

- mock-ups of various components of the plant.

▲ *see Glossary*

 **A4** European Activity

European activity is in the hands of a number of institutions and committees. Discerning the hierarchy is not evident but the European Strategic Energy Technology (SET)-Plan aims to transform energy production and use in the European Union with the goal of achieving worldwide leadership. The implementation mechanisms of the SET-Plan are the European Industrial Initiatives (EII) and the European Energy Research Alliance (EERA). The Strategic Energy Technologies Information System Information for Decision-making (SETIS) is led and managed by the Joint Research Council's Institute for Energy and Transport. It covers most aspects of the energy sector. In the coming decades, it is claimed, research will focus mainly on development of sustainable nuclear energy through fast (Gen IV) reactors of which the sodium cooled fast prototype reactor (ASTRID) is highly ranked and intended to supply the grid with 600 MWe. The Lead-cooled Fast Reactor (LFR) and Gas-cooled fast reactor (GFR) are smaller, and commercial deployment is expected ten years later. These particular Gen IV options are currently expected to have 20–30% higher investment cost than Gen III reactors. The MSR option is not considered in this scenario.

Since 2001 collaboration between European and Russian institutions, supported by Euratom and Rosatom grants, succeeded in creating research tools and generating data on molten salt reactors. Research and development of Molten Salt Fast Reactor (MSFR) concepts was started about ten years ago, led by the National Center for Scientific Research in France (CNRS), a research organization dealing more with science than engineering, thus explaining why up to now, in Europe, only basic studies have been conducted. This approach continues for the immediate future. Current plans include evaluation of safety and optimization of design in terms of inherent MSFR characteristics before more investment in design and engineering is seen. MSFR safety is a present target for the Horizon 2020 Safety Assessment of the MSFR project (SAMOFAR) that will last from August 2015 to July 2019. This is now led by a team at the Technical University of Delft in The Netherlands.

France

In addition to the Euratom and Rosatom projects, French institutions also support academic activities on the MSFR in an interdisciplinary research program named NEEDS (Nucléaire, Energie, Environnement, Déchets, Société) which promotes nuclear research taking account of nature and society. French MSR activity is centred at CNRS Grenoble.

Recent work comprised evaluation of deployment capacities by simulations of MSFR configurations corresponding to three types of initial liquid fuel composition demonstrating the high flexibility of the concept. An experimental circulation loop has been tested in Grenoble to study gas injection and separation in liquid fluorides. An 80 litre *FLiNaK*▲ loop operates at 600°C with flow rate of about 2 litres per second. It is equipped with ultrasound velocimeter, gas pressure regulation and salt level controls, freeze plug and mechanical ball valve. A chemically altered water mock-up version shows that a bubble-salt separation rate in excess of 95% can be reached.

An innovative code simulating transients by coupling neutronics and thermal hydraulics has been developed. These preliminary simulations confirm good behaviour of the MSFR for normal and accidental transient calculations such as load following and reactivity insertion.
References for further information on the MSFR and works at CNRS are available at [i, ii, iii, iv, v, vi].

▲ *see Glossary*

The Netherlands

Delft University of Technology (TU Delft)

The research team at Delft advocates the implementation of thorium molten salt reactors. They specialise in neutronics modelling of advanced reactor configurations and safety analysis. They were a partner in the early MARS and EVOL projects but have now taken a leadership role overseeing the next MSR project supported by the European Commission through Euratom, Safety Assessment of the Molten Salt Fast Reactor (SAMOFAR). The consortium across Europe will carry out experiments and safety analyses in the following areas:

Experimental work

- Salt conditions at walls of piping and freeze plug
- Optimal geometry of the freeze plug
- Natural circulation dynamics of salt
- Demonstrate extraction processes

Safety assessment

- Measuring safety related data of salts
- New integral safety approach
- Software simulator (start-up, load-following, …)
- Transient scenarios (multi-physics, uncertainty, …)
- Safety of chemical processes

Germany

The Institute for Transuranium Elements (ITU)

Based in Karlsruhe, Germany, this institute is one of seven led by the Joint Research Centre, a Directorate-General of the European Commission. There are about 300 staff with access to an extensive range of advanced facilities, many unavailable elsewhere in Europe, including substantial infrastructure for working on MSR fuel and clean-up technology. The institute has extensive experimental capability for preparing and studying actinide fluoride salts (Th, U, Pu) and beryllium.

Institute facilities include:

- A fluorination line (using HF) to convert oxides to fluorides, and to purify salt.
- Calorimetric equipment to determine melting temperatures, heat capacities, enthalpies of fusion and phase diagrams of salt mixtures.
- Mass spectrometric equipment to measure vapour pressure and thermodynamic activities of salt mixtures. The lead shielded facility allows experimental determination of fission product release of irradiated salt samples.
- High temperature Raman spectroscopy of molten salt mixtures to analyse the chemical speciation.
- Electrochemical equipment for measuring electro-potentials of metal/fluoride couples and testing the electrochemical separation processes.

This institute has been active in the MSFR programmes. The experimental work that focussed on improving the thermodynamic description of the $LiF-ThF_4-BeF_2-UF_4$ system has led to the compilation of an extensive thermodynamic database that describes properties such as heat capacity, melting, vapour pressure, boiling point and solubility of actinide fluorides. A substantial effort has been made to optimise the purity of the salt components, and high purity ThF_4 and UF_4 have been obtained, and the synthesis of PuF_3 is ongoing.

Other work focusses on the behaviour of fission products in the salt, aiming at demonstrating the retention capability of the salt for fission products that are volatile in LWR fuels. Thorium salt is being prepared for an irradiation experiment in the High Flux Reactor in Petten, in a joint effort with NRG using the reactor at Petten in the Netherlands.

Dual Fluid Reactor

Developed at the Institute for Solid-State Nuclear Physics, Berlin, Germany, this is not a traditional MSR. It operates in the fast spectrum with two fluids going through the reactor core. The fuel salt flows slowly and the coolant salt at a faster rate to maximise heat transfer. Two versions are proposed, the DFR/s which utilizes an undiluted actinide chloride-37 salt, and the DFR/m featuring a liquid actinide metal alloy - both being cooled by liquid lead. It has an operating temperature around 1000°C.

More information: www.dual-fluid-reactor.org

Denmark

Copenhagen Atomics

With a goal to develop MSR technology generally, Copenhagen Atomics advocates a thorium reactor configured as a burner of spent nuclear fuel. The reactor is fed by nuclear waste which is reprocessed on site by a reprocessing module. It is designed for factory line production and easy transportation with the possibility for centrally decommissioning. They focus on chemical reprocessing technology and measurement and control systems as oppose to reactor design. They intend to run radioactive salt experiments in Research Center Řež in Czech Republic and are actively seeking partnerships to advance this.

Russia

MOSART

The first molten salt work in Russia occurred in the mid-1970s. Over the past decade work has recommenced based out of the Kurchatov Institute in Russia working on MOlten Salt Actinide Recycler & Transmuter (MOSART). Early work concentrated on a single fluid burner which progressed to a two fluid fast breeder configuration. The system is initially fuelled with compositions of plutonium plus minor actinide trifluorides from PWR spent fuel without U-Th support. As time progresses the thorium blanket salt provides more of the fissile fuel as U-233. FLiBe salt is proposed operating at temperatures above 700°C. The team have participated closely with the European MSFR project. More information is available at reference[vii].

5 | Japanese MSR Activity

Molten Salt Reactor technology was tirelessly promoted in Japan by Professor Kazuo Furukawa (deceased 2011) for the decades since he had finished working for Alvin Weinberg at Oak Ridge National Laboratory in the 1970s. His group in Japan have been investigating the graphite moderated MSR "FUJI" and related accelerator technology since the 1980s until now[viii, ix]. FUJI is mostly based on the MSBR design with several improvements, building on ORNL development from 1950s to 1970s. This group have provided various design results, such as detailed concept design of the self-sustaining FUJI-U3, FUJI-Pu with plutonium start-up and transmutation capability of minor actinides, super-FUJI at 1,000 MWe and a pilot plant mini-FUJI.

In order to promote MSR development, Furukawa and his group established the International Thorium Molten-Salt Forum (ITMSF) in 2008, and the Thorium Tech Solution Inc. (TTS) in 2010. ITMSF is a Non-Profit-Organization, which is composed of researchers and engineers among 13 countries who are involved or interested in MSRs and related thorium cycles. So far, 17 seminars have been held in Japan. ITMSF is an observer member of the MSR subgroup of the Generation IV International Forum.

Under the Atomic Energy Society of Japan, a special committee was started known as "Nuclear Application of Molten Salt". Experts including several ITMSF members are discussing MSRs, dry-processing, fusion application, and so on. A working group in this committee is participating in the ANS20.1 committee, where General Design Criteria on the *FHR*▲ is discussed.

TTS's final target is to commercialise the FUJI based on the developments within ITMSF. TTS is now working on molten salt technology at the test reactor at Halden, Norway. Irradiation experiments are planned in the near future. Kyoto Neutronics is designing a small thorium MSR with integrated heat-exchangers, named UNOMI (Universally Operable Molten-salt reactor Integrated)[x]. There is Japanese dedication to the development of MSR technology and it will continue in the future.

▲ *see Glossary*

Appendix A - References

i   M. Brovchenko et al, *Design-related Studies for the Preliminary Safety Assessment of the Molten Salt Fast Reactor*, Nuclear Science and Engineering, 175, 329–33 (2013)

ii   D. Heuer et al, *Towards the Thorium Fuel Cycle with Molten Salt Fast Reactors*, Annals of Nuclear Energy 64, 421–429 (2014)

iii A. Laureau et al, *Coupled Neutronics and Thermal-hydraulics Transient Calculations based on a Fission Matrix Approach: M&C*, SNA and MC Method International Conference, Nashville, USA (2015)

iv   E.Merle-Lucotte et al. *Physical assessment of the load following and starting procedures for the molten salt fast reactor*, ICAPP 2015 International Conference, Nice, France (2015)

v   IRSN report: *Review of Generation IV Nuclear Energy Systems*, April 2015

vi   E.Merle-Lucotte, *The concept of Fast Spectrum Molten Salt Reactor (MSFR)*, Presentation at French-Swedish Seminar on Future Nuclear Systems - KTH – Stockholm, Sweden. December 2013

vii V. Ignatiev, O. Feynberg, *Molten Salt Reactor: Overview and Perspectives*, Proceedings of OECD Actinide and Fission Product Partitioning and Transmutation 11th Information Exchange Meeting (2012)

viii K. Furukawa et al., *A Road Map for the Realization of Global-scale Thorium Breeding Fuel Cycle by Single Molten-fluoride Flow*, Energy Conversion and Management, vol.49, p.1832-1847 (2008).

ix   R. Yoshioka, *Nuclear Energy Based on Thorium Molten Salt, chapter-23 of the book Molten Salts Chemistry: From Lab to Applications, edited by F. Lantelme and H. Groult*, Elsevier Inc., USA (2013).

x   T. Kamei, *Review of R&D of Thorium Molten-Salt Reactor*, Nuclear Safety and Simulation, Vol.4, No.2, P.80-96 (2013)

Energy Process Developments Ltd.

Appendix B

Early MSR Activity

Liquid Fuel Experiments in the Manhattan Project

Groundwork for molten salt reactors was started with the secret Manhattan Project using aqueous homogeneous reactors (AHR) with soluble nuclear salts (usually uranium sulphate or uranium nitrate) dissolved in water mixed with the coolant and moderator. They tended to fail on account of corrosion and radiolysis causing gas bubbling. Objectives included creating weapons material and providing for small power plants for military use.

Information is available in Chapter 5 of the Final Report of the Los Alamos Document Retrieval and Assessment (LAHDRA) Project 2009.

Projects included:

- LOPO. LOw POwer. Used U-235 as uranyl sulphate.

- HYPO. HIgh Power. Used uranyl nitrate. Ran at 5.5kW until 1951.

- SUPO. Super Power. Larger than HYPO, assisted the weapons program until 1974.

Further liquid-fuelled experimental work continued with the Los Alamos Power Reactor Experiment (LAPRE) tasked with developing high power density reactors:

- LAPRE I (1955). Used uranium oxide dissolved in phosphoric acid. Corrosion problems halted it despite gold plating.

- LAPRE II (1956-59). Used a different fuel solution but corrosion problems continued and project was dropped in 1960.

- LAMPRE (1961-63). Used molten plutonium cooled by molten sodium. Ran for 20,000 hours.

Aircraft Reactor Experiment (ARE)

There have been just three molten salt reactors operated at or above criticality. The first was the 2.5 MWth aircraft reactor experiment in 1954 as part of a programme to develop a nuclear powered aircraft. The second was in 1957, the Pratt and Whitney Aircraft Reactor-1 (PWAR-1) and operated for a few weeks essentially as a zero power reactor; reports are that it was installed for test flights. The third and last was the molten salt experiment reactor (MSRE) that went critical in 1964. All three were housed in the Critical Experiments Facility of the Oak Ridge National Laboratory, a building that still survives.

The first, the ARE, used molten fluoride salt NaF-ZrF_4-UF_4 (53-41-6 mol%) as fuel. It was moderated by beryllium oxide, employed liquid sodium as secondary coolant and reached 860°C. It operated for 100 MW-hours over nine days in 1954. The image on page 18 shows the reactor undergoing tests suspended in mid air. An aircraft nuclear power unit project was shown to be impracticable and was abandoned.

Molten Salt Reactor Experiment (MSRE)

The MSRE ran successfully at ORNL during 1965 – 1969, operating for the equivalent of one and a half years at full power. It was a prototype for subsequent liquid thorium reactor (LFTR) proposals but with the breeding blanket of thorium salts omitted to facilitate neutron measurement. This 7.4 MWth single fluid reactor operated in the thermal spectrum moderated by pyrolytic graphite. Fuel salt for the MSRE was LiF-BeF_2-ZrF_4-UF_4 (65-29-5-1), first with U-235 later with U-233 as primary fissile fuel. This reactor, with the salt in solid state underground, remains at Oak Ridge under care and maintenance. It is ready to dismantle. The reactor site was visited as part of this project.

This experimental reactor technology is developed in most MSR design proposals including those assessed in the present report. Reasons given at the time for discontinuing an MSR programme were that a consolidated research agenda was needed for more advanced development projects. US priority had been given to plutonium production during the cold war.

The MSRE building as it stands today visited by EPD directors, Jasper Tomlinson and Rory O'Sullivan

Proposed Molten Salt Breeder Reactor (MSBR)

Oak Ridge National Laboratory prepared detailed concept designs for a two fluid thorium breeder reactor and a single fluid breeder reactor in the 1970s. This was well documented and several MSR designs today have used these principles. Work in the USA on MSR development was brought to a halt in 1976 because of other government priorities.

Proposed Molten Salt Epithermal (MOSEL) Reactor

A programme in Germany which omitted graphite in the reactor core to harden the spectrum and reach into the peak region of the U-233 neutron yield in the epithermal spectrum. They furthered research on NaF-ZrF$_4$ salts from aircraft experiment along with KF salts with U/Th tetrafluorides. ORNL and General Electric Company were involved in this programme.

Proposed Molten Salt Fast Reactor (MSFR)

This was a secret UK Atomic Energy Agency detailed concept design completed in 1976 for a helium cooled 2.5 GWe molten salt fast spectrum reactor with plutonium fuel in a chloride salt. The proposal included the possibility of modular construction. It was considered ready for implementation. Various coolants such as lead were assessed. It was more complicated than fast spectrum proposals today since the emphasis, at that time, on breeding was a response to mistaken perceptions of a limit to uranium resources for fuel.

A full design report was shared with France in an attempt to promote collaboration. The project was finally shelved, most likely in 1976, the same year as the design report was completed. The full report has rested undisturbed in the UK National Archives until very recently.

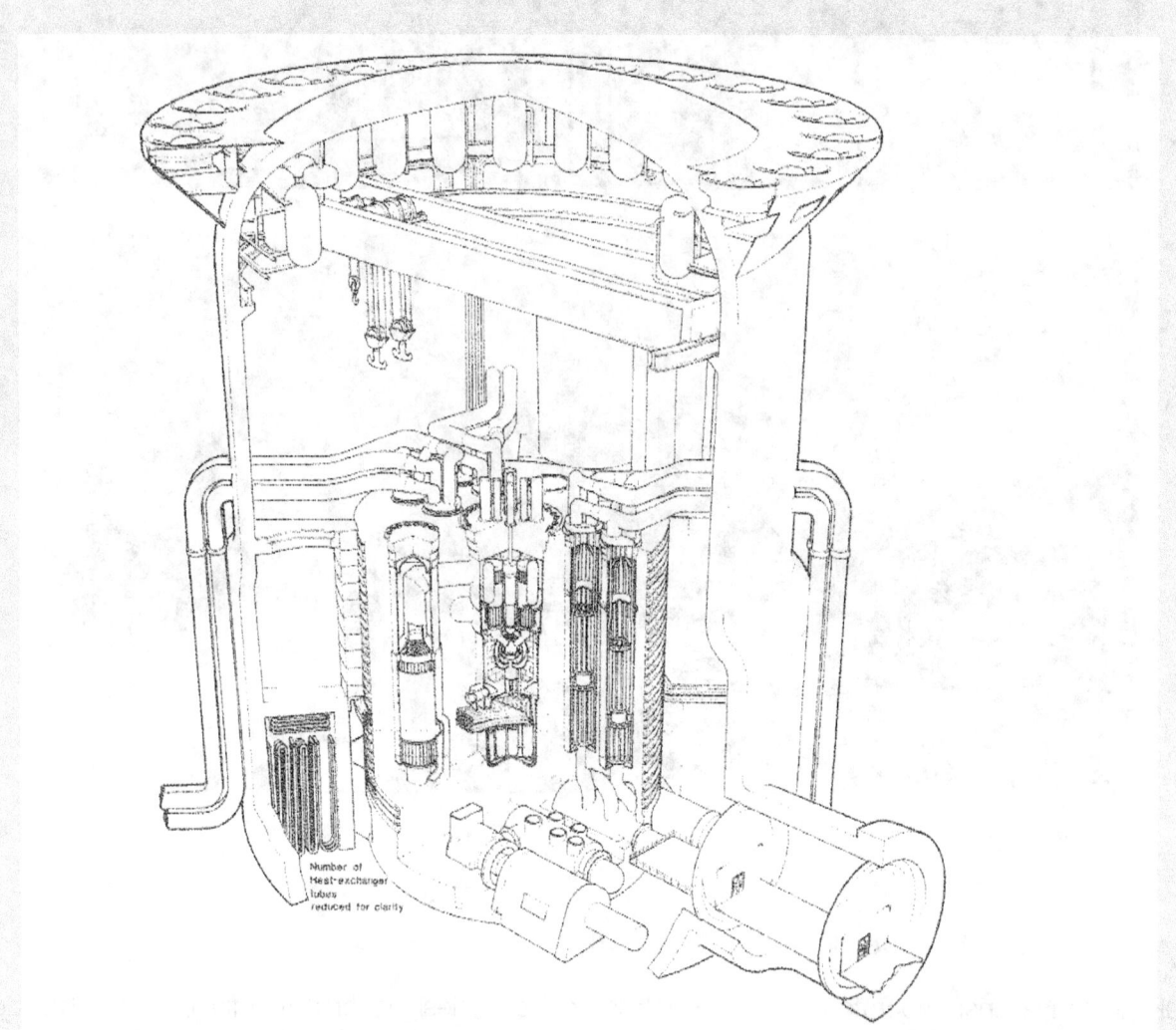

The helium-cooled MSFR with integrated 4 set double shaft gas turbine within a prestressed concrete vessel (J.Smith, W.E,Simons. AEEW-R956, August 1976). Image scanned from original documents at the National Archives

Appendix C

Public Opinion Poll

EPD Opinion Poll on Advanced Nuclear

Fieldwork was conducted using Ipsos MORI's i:Omnibus service among an online quota sample of adults aged 16-75 in Great Britain (n = 1007) between 27th and 31st March 2015. Data were weighted by age, gender, region, working status and social grade to the known online population profile. This online sample automatically excludes anyone who is not comfortable with the use of the internet, or who has no access to it. There is the possibility that such people would be more negative towards nuclear energy.

In summary, the results are more positive towards nuclear energy than has been seen by Ipsos MORI for a few years - though they never asked these exact questions. A notable finding at Q1 was that more selected the "challenge old technology" option than the "known technology" option. A surprising 23% claim that they have heard of Generation IV reactors as a better source of nuclear energy.

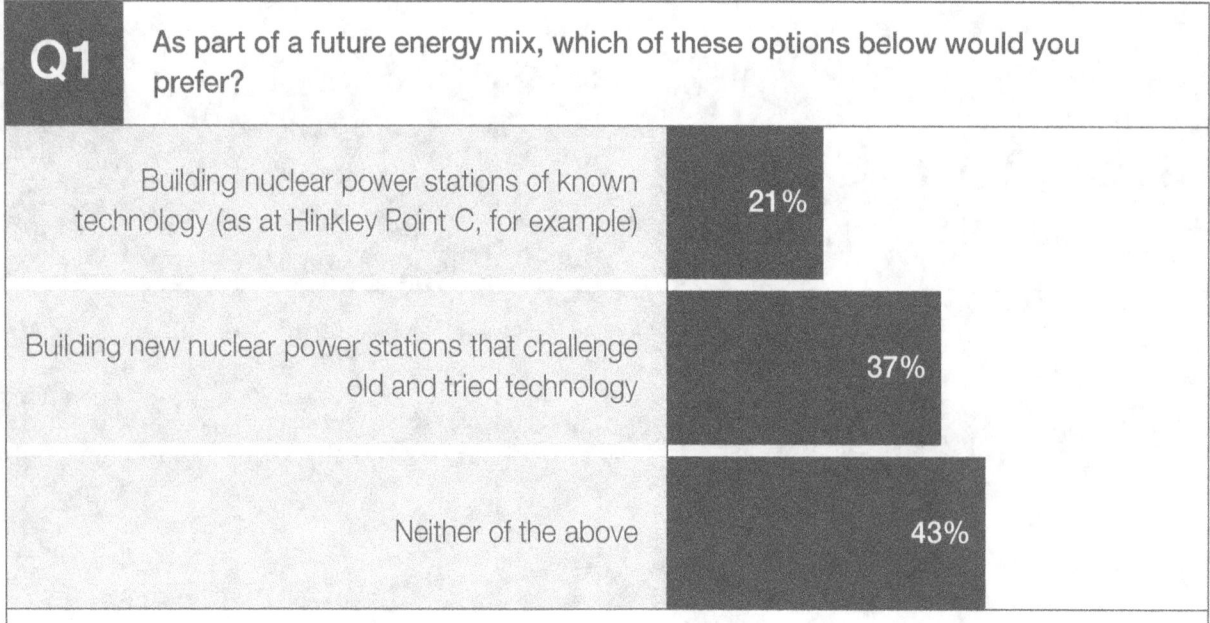

Q1 As part of a future energy mix, which of these options below would you prefer?

Building nuclear power stations of known technology (as at Hinkley Point C, for example)	21%
Building new nuclear power stations that challenge old and tried technology	37%
Neither of the above	43%

Note 1

In summary there are 57% for more nuclear, 43% against. Taking only the respondents who responded to Q2 as in favour of the expansion of nuclear power (who comprise 41% of the total), 58% of them would prefer to see new technology that challenges existing.

Q2 — To what extent do you agree or disagree with each of these statements about nuclear power?

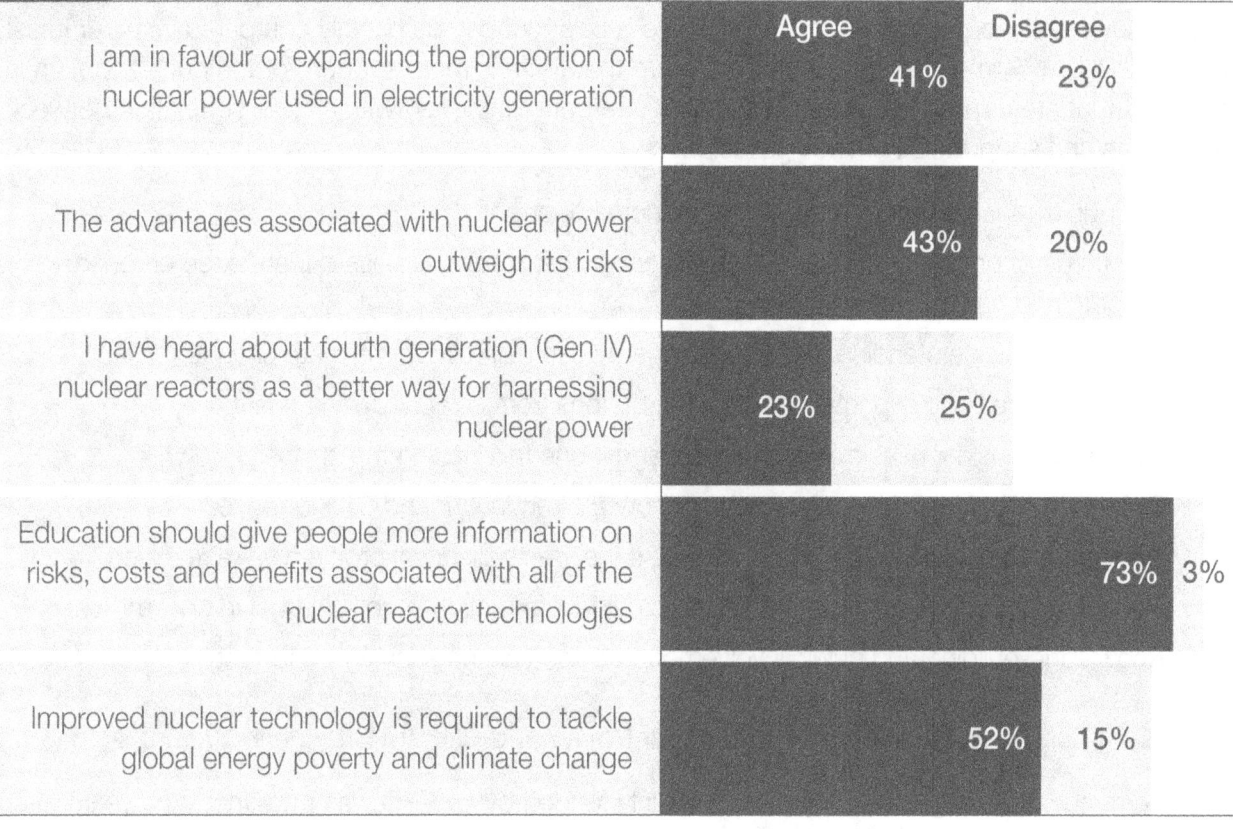

	Agree	Disagree
I am in favour of expanding the proportion of nuclear power used in electricity generation	41%	23%
The advantages associated with nuclear power outweigh its risks	43%	20%
I have heard about fourth generation (Gen IV) nuclear reactors as a better way for harnessing nuclear power	23%	25%
Education should give people more information on risks, costs and benefits associated with all of the nuclear reactor technologies	73%	3%
Improved nuclear technology is required to tackle global energy poverty and climate change	52%	15%

Note 2

The remaining percentage of respondents answered 'Neither Agree nor Disagree'.
As seen in previous studies by Ipsos MORIs, men and older people were seen to be more favourable towards nuclear energy in general compared with women or younger people.

Q3 — Which, if any, of these issues below are most important to consider before building new nuclear power stations?

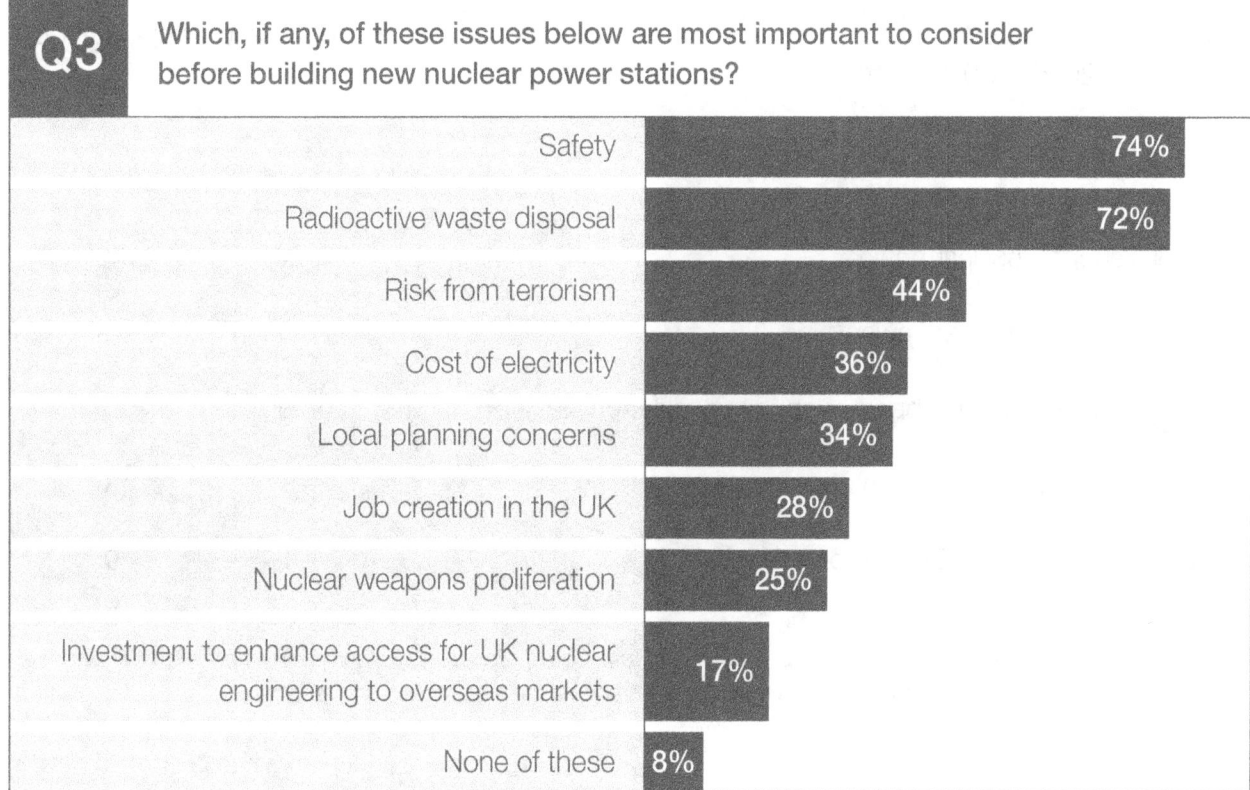

Safety	74%
Radioactive waste disposal	72%
Risk from terrorism	44%
Cost of electricity	36%
Local planning concerns	34%
Job creation in the UK	28%
Nuclear weapons proliferation	25%
Investment to enhance access for UK nuclear engineering to overseas markets	17%
None of these	8%

OECD Report on Public Attitudes to Nuclear Power, 2010

The OECD published a 2010 report on Public Attitudes to Nuclear Power. This 54 page document gives many results for European countries. There is little that can be directly compared with our latest Ipsos MORI poll. Some relevant conclusions were that there is more support for the nuclear option in countries that are using it and these populations are better informed with a clear correlation between knowledge and support. Other findings included:

- The public regards many other issues as more important
- Two thirds of people understand that nuclear power makes for less dependence on energy imports; and half that it helps keep prices down
- There is little understanding that nuclear energy can help combat climate change
- There are unrealistic expectations for contributions from solar and wind energy
- Half reckon nuclear risks outweigh advantages, one third taking the opposite view
- Opinion in countries with nuclear power show 28% in favour and 31% opposed
- Opposition to nuclear power would reduce if the waste disposal problem were resolved
- Scientists and NGOs are most trusted to provide information, with little trust in governments
- Knowledge-building and public involvement are seen as increasingly important

UKERC Report on Public Attitudes to Nuclear Power and Climate Change in Britain, 2013

This more recent 38-page report of a study mainly funded by UKERC was published two years after the Fukushima accident. A significant conclusion was that nuclear power remains one of the least favoured energy sources. This report describes the main findings of the British survey conducted in March 2013. The 38 questions of the survey were added to Ipsos MORI's face-to-face omnibus that was conducted between 8 and 26 March 2013. The survey included a range of assessments relating to how the general public thinks about nuclear power. About the same number of people generally either supported (32 per cent) or opposed (29 per cent) nuclear power in 2013.

A tabulation of answers concerning nuclear power (in %) can be summarized to show how attitudes shifted from 2005 to 2013:

	2005	2013
Overall, I support nuclear power	26	32
Overall, I oppose nuclear power	37	29
I am not sure whether I support or oppose nuclear power	32	27
I don't care what happens with nuclear power	3	3
Other / None of these / Don't know	1	9

The main authors of this 2013 report were psychologists from Cardiff University. Rather than simply report findings, a determination to uncover thoughts and feelings can be seen amongst the many tables and graphs.

Appendix D

Table of Nuclear Licensed Sites in the UK

Site	Owner	Operator	Location
Aldermaston & Burghfield	Crown	AWE	Nr Reading
Amersham Laboratories	GE Healthcare	GE Healthcare	Amersham Bucks
Barrow	BAE Systems		Barrow-in- Furness, Cumbria
Berkeley	NDA	Magnox Limited	Gloucestershire
Bradwell	NDA	Magnox Limited	Bradwell-onSea, Essex
Capenhurst	URENCO	URENCO UK Limited	Capenhurst, Chester
Cardiff Laboratories	GE Healthcare	GE Healthcare	Cardiff
Chapelcross	NDA	Magnox Limited	Annan, Dumfrieshire
Consort Reactor	Imperial College	Imperial College	Ascot, Surrey
Devonport Dockyard	Devonport Royal Dockyard	Devonport Royal Dockyard	Devon coast
Dounreay	NDA	RSRL	Thurso, Caithness
Dungeness A	NDA	Magnox Limited	Kent Coast
Dungeness B	Edf	Edf	Kent Coast
Hartlepool	Edf	Edf	Hartlepool
Harwell	UKAEA	RSRL	Didcot, Oxfordshire
Heysham 1	Edf	Edf	Lancashire Coast
Heysham 2	Edf	Edf	Lancashire Coast
Hinkley A	NDA	Magnox	N Somerset Coast
Hinkley B	Edf	Edf	N Somerset Coast
Hinkley C	Edf	NNB Generation	N Somerset Coast

Energy Process Developments Ltd.

Status	Comments
2 adjacent Defence Sites	Security considerations and defence-focussed
Isotope services	Focussed on healthcare activities. Nr centre of Amersham
Submarine reactor assembly/ construction	Defence Focussed + commercial priorities
Decommissioning Magnox Reactor	Station to enter Care and Maintenance 2021
Decommissioning Magnox Reactor	Enters care & maintenance 2016/17
Decommissioning/enrichment services	Enrichment/decommissioning.
Healthcare activities	Focussed on healthcare activities. Site being shrunk and activities reduced
Decommissioning Magnox Reactor	Interim care & maintenance.Scottish Govt opposes new nuclear build
Decommissioning Research Reactor	Undergoing decommisioning
Submarine support	Security considerations and defence-focussed
Decommissioning Site	History of prototype reactors - Dounreay Fast Reactor and Prototype Fast Reactor from 1957 - early 90s. Geographically isolated on Northern tip of Scotland near Thurso. Scottish govt policy opposes new build and has vires over environmental regulation. The NDA has publically stated it intends Dounreay to be 'fuel free' by 2029.
Decommissioning Magnox Reactor	Decommissioning. To enter interim care & maintenance 2019-2023
Operating AGR	-
Operating AGR	Nominal shutdown date 2024
Undergoing decommissioning - and being delicensed	Situated nr Didcot in Oxfordshire. Harwell was home to a number of experimental reactors which have all been closed down. The strategy for Harwell is to decommission and clear the site within around 5 years and sell the land which because of its location, would be valuable. As land becomes cleared it is being delicensed piecemeal.
Operating AGR	Commercial generating priorities
Operating AGR	Commercial generating priorities
Decommissioning magnox Reactor	EPR Planned on adjacent site. Plant entering Care and Maintenance.
Operational AGR	EPR Planned on adjacent site
New Reactor	EPR Planned

Site	Owner	Operator	Location
Hunterston A	Magnox Ltd	Magnox	W Kilbride
Hunterston B	Edf	Edf	W Kilbride
LLWR Repository	NDA	LLWR	Drigg near Sellafield
Metals Recycling Facility	Studsvik	Studsvik	Workington, Cumbria
Oldbury	NDA	Magnox Limited	Gloucestershire
Raynesway	Rolls Royce	Rolls Royce	Derby
Rosyth Dockyard	Rosyth Royal Dockyard Ltd	Babcock Marine	Edinburgh
Sellafield	NDA	Sellafield Ltd	W Cumbria
Sizewell A	NDA	Magnox	Leiston,Suffolk
Sizewell B	Edf	Edf	Adjacent to Sizewell A
Springfields Works	Springfields Fuels	Springfields Fuels	Somerset - W Coast
Torness	Edf	Edf	Dunblane
Trawsfynydd	NDA	Magnox Limited	N Wales
Winfrith	NDA	Magnox Limited	Dorset
Wylfa	NDA	Magnox Limited	Anglesey

Status	Comments
Decommissioning Magnox Station	Decommissioning & demolition. Enters care & maintenance 2022. Scottish Govt opposes new nuclear build.
Operational AGR	Commercial generating priorities + Scottish Govt opposes new nuclear build
Low level waste shallow burial	Operating purely as a disposal site.
Waste Services	Subsidiary of Swedish waste management company
Decommissioning Magnox Reactor	Defuelling. Enters care & maintenance 2027. Horizon (owned by Hitachi) plan new station at Oldbury.
2 adjacent Defence Sites	Defence focussed + commercial priorities
Submarine support	Security considerations and defence-focussed
Reprocessing and decommisioning	Multiple plant complex.Three new reactors possible at nearby Moorside by NUGEN. Spent fuel and plutonium stockpile and adjacent reprocessing facilities would be beneficial.
Decommissioning Magnox Reactor	Plant shutdown and defuelling completed Aug 2014. Magnox plans for Care & Maintenance for about 80 years before decommissioning of the reactor itself.
Operating PWR	An area of land adjacent to Sizewell B was earmarked for a new reactor ('Sizewell C')
Fuel Manfacture and Decommissioning	Post Operational Clean-out and decommissioning.Parent Body is Westinghouse
Operational AGR	Scottish Govt opposes new nuclear build
Decommissioning Magnox Reactor	Decommissioning and demolition. Enters care & maintenance 2016/7
Decommissioning	Situated near Wool in Dorset. Some of the site has been delicensed and now forms an adjoining business park. There are two reactors in the process of decommissioning (Dragon and SGHWR) and the aim is to achieve clearance so that the site can be released as heathland.
New reactor likely for site	Horizon (owned by Hitachi) plan new station at Wylfa. Defuelling. Reactor 2 shutdown 2012. Reactor 1 due to shutdown end of 2015.

Appendix E

MSRs from a Nuclear Insurer's Perspective

JLT's Global Nuclear Industry Practice Group with SIACI SAINT HONORE (S2H), brings together London's leading construction and private finance expertise and the leading nuclear insurance broker in Europe to help those involved with the decommissioning, construction and operation of nuclear power facilities to develop innovative insurance and risk solutions. They have provided insight on the insurability of a prototype molten salt reactor.

Molten Salt Reactor technology is well established, and has been proven to be a safer and more cost effective use of nuclear fission for power generation purposes. However, newer models of molten salt reactors, which have yet to be built on a commercial scale, would still be considered "prototypes" from an insurance perspective. An additional constraint towards insuring these new models would be overcoming the nuclear insurance markets inherent preference for light water reactors, due to their familiarity of the technology, decades of experience, and loss data established over time. A final hurdle to overcome is that regardless of the elimination of most potential causes of a catastrophic nuclear incident, MSRs would still be beholden to adhere to the international conventions for civil nuclear liability as well as domestic legislation based on these conventions, which were designed and developed with light water reactors in mind and the potential risks that these reactors pose. Let's examine each of these in greater detail:

Prototype Status
Any new reactor design is subject to regulatory approval which is an arduous and lengthy process; MSRs are not immune to this process. When regulatory approval is granted, until there is significant run time or operational experience of a certain reactor design, it will be considered by the insurance industry as a prototype. The impact of this rather subjective designation varies from market to market, but in general the effect is a reduction in insurance coverage (compared to that of light water reactors), lower levels of capacity, and a much more arduous underwriting process.

For example, as a prototype reactor, MSRs would not be able to obtain full defect liability cover that is available to equipment with substantial run times. Typically, most policies allow for LEG (London Engineering Group) 3 cover, which provides cover for resulting damages from a defect in design, materials and/or workmanship, along with coverage to replace the defective part. LEG 2 cover, a step below, provides only cover for resulting damages but not for the defective part. A prototype design would not be able to obtain LEG 3 cover and would be subject to LEG 2 cover at best. This is one example but other clauses for extensions in cover and the breadth of cover would also be reduced for prototypes.

The prototype status also means that fewer markets would be willing to add this risk to their existing portfolio. Insurance markets tend to be risk averse, and many markets will not write what they do not understand or risks that fall into easily identifiable categories. Those that would write the cover may limit the amount of their net capacity to provide for the risk, until they become more comfortable with the exposure and have earned premium over time.

Prototypes are also subject to greater scrutiny from an underwriting perspective. Technical reports, engineering studies, detailed loss scenarios and other documentation would be required by underwriters to pore over so that they have full appreciation for the inherent risks and the likelihood of a claim occurring.

Prototypes are also generally subject to higher premiums as they are viewed as having a higher risk exposure due to the unavailability of documented loss history.

Preference for What You Know

The nuclear insurance market is far different from the commercial insurance market. This niche market is dominated by domestic nuclear pools, who underwrite nuclear risks on behalf of their member insurers who provide their net nuclear capacity to the pool. These domestic nuclear pools then reinsure each other to increase their available capacity to provide cover for large nuclear risks, such as commercial nuclear power plants. The nuclear pooling system has been providing cover and issuing policies for decades based on the dominant technology in the sector – light water reactors. Their approach to these risks tend to be rather standardised, and an MSR design would not fit into this standardised approach due to the differences in many phases of the technology. It is quite possible that the pools would prefer to simply steer clear of providing coverage for a non-standard design, and reserve their capacity for other light water reactors (in other words, "write what they know").

As MSRs become more commonplace over time – the goal of this technology – nuclear pools will ultimately need from a commercial perspective to write this cover. However, the nuclear pools are the only option to provide high levels of capacity for nuclear risks and operate on somewhat of a monopoly basis. Therefore if they do offer cover for MSR designs, they would likely charge higher premiums than for light water reactors, and operators would have no real alternatives.

Paris & Vienna Conventions

These conventions for civil nuclear liability were developed and written by nuclear operators, for nuclear operators. The nuclear operators of the time used light water reactor technology, albeit earlier generations. Because of the nature of nuclear incidents with light water reactors, limits and requirements were designed with these factors in mind. MSRs due to their inherent safety of lower operating pressures (atmospheric), releases of pressure building gasses, and containment of fuel and coolant salts, plus the inherent stability of the salts themselves, are not subject to the same potential catastrophic incidents which could cause public harm. However, as the MSR technology still involves nuclear fission, they would still be obligated to adhere to the international conventions and the domestic civil nuclear legislation that mirrors these conventions. If we take for example the revisions of the Paris Convention from 2004, which are due to come in force in the very near future, operators would be required to provide €1.2 billion in nuclear liability coverage in the UK. There are other aspects to the revision that are currently uninsurable. Even though MSRs do not have the same risk profile as light water reactors, operators would still be required to take out insurance in these amounts in order to maintain their operating license, and the uninsurable heads of cover would have to be retained on their balance sheets.

185 New Kent Road
London
SE1 4AG
UK

For further information please contact:
info@energyprocessdevelopments.com

Jasper Tomlinson for policy issues:
jasper@energyprocessdevelopments.com
0207 3479457

Rory O'Sullivan for engineering details:
rory@energyprocessdevelopments.com

Trevor Griffiths for molten salt chemistry:
t.r.griffiths@gmail.com

www.ingramcontent.com/pod-product-compliance
Lightning Source LLC
Chambersburg PA
CBHW081049170526
45158CB00006B/1917